Critical Criminological Perspectives

The Palgrave *Critical Criminological Perspectives* book series aims to showcase the importance of critical criminological thinking when examining problems of crime, social harm and criminal and social justice. Critical perspectives have been instrumental in creating new research agendas and areas of criminological interest. By challenging state-defined concepts of crime and rejecting positive analyses of criminality, critical criminological approaches continually push the boundaries and scope of criminology, creating new areas of focus and developing new ways of thinking about, and responding to, issues of social concern at local, national and global levels. Recent years have witnessed a flourishing of critical criminological narratives and this series seeks to capture the original and innovative ways that these discourses are engaging with contemporary issues of crime and justice.

Series editors:
Professor Reece Walters
Faculty of Law, Queensland University of Technology, Australia
Dr Deborah Drake
Department of Social Policy and Criminology, The Open University, UK

Titles include:

Kerry Carrington, Matthew Ball, Erin O'Brien and Juan Tauri (*editors*)
CRIME, JUSTICE AND SOCIAL DEMOCRACY
International Perspectives

Claire Cohen
MALE RAPE IS A FEMINIST ISSUE
Feminism, Governmentality and Male Rape

Marianne Colbran
MEDIA REPRESENTATIONS OF POLICE AND CRIME
Shaping the Police Television Drama

Pamela Davies, Peter Francis and Tanya Wyatt (*editors*)
INVISIBLE CRIMES AND SOCIAL HARMS

Melissa Dearey
MAKING SENSE OF EVIL
An Interdisciplinary Approach

Michael Dellwing, Joseph A. Kotarba and Nathan W. Pino (*editors*)
THE DEATH AND RESURRECTION OF DEVIANCE
Current Ideas and Research

Deborah Drake
PRISONS, PUNISHMENT AND THE PURSUIT OF SECURITY

Matt Long and Roger Hopkins Burke
VANDALISM AND ANTI-SOCIAL BEHAVIOUR

Margaret Malloch and William Munro (*editors*)
CRIME, CRITIQUE AND UTOPIA

Erin O'Brien, Sharon Hayes and Belinda Carpenter
THE POLITICS OF SEX TRAFFICKING
A Moral Geography

Maggie O'Neill and Lizzie Seal (*editors*)
TRANSGRESSIVE IMAGINATIONS
Crime, Deviance and Culture

Ragnhild Aslaug Sollund (*editor*)
GREEN HARMS AND CRIMES
Critical Criminology in a Changing World

Diane Westerhuis, Reece Walters and Tanya Wyatt (*editors*)
EMERGING ISSUES IN GREEN CRIMINOLOGY
Exploring Power, Justice and Harm

Tanya Wyatt
WILDLIFE TRAFFICKING
A Deconstruction of the Crime, the Victims, and the Offenders

---

Critical Criminological Perspectives
Series Standing Order ISBN 978–0–230–36045–7 hardback
(*outside North America only*)

You can receive future titles in this series as they are published by placing a standing order. Please contact your bookseller or, in case of difficulty, write to us at the address below with your name and address, the title of the series and the ISBN quoted above.

Customer Services Department, Macmillan Distribution Ltd, Houndmills, Basingstoke, Hampshire RG21 6XS, England

---

Green Harms and Crimes

Critical Criminology in a Changing World

Edited by

Ragnhild Aslaug Sollund
University of Oslo, Norway

First published 2015 by
PALGRAVE MACMILLAN

Palgrave Macmillan in the UK is an imprint of Macmillan Publishers Limited,
registered in England, company number 785998, of Houndmills, Basingstoke,
Hampshire RG21 6XS.

Palgrave Macmillan in the US is a division of St Martin's Press LLC,
175 Fifth Avenue, New York, NY 10010.

Palgrave Macmillan is the global academic imprint of the above companies
and has companies and representatives throughout the world.

Palgrave® and Macmillan® are registered trademarks in the United States,
the United Kingdom, Europe and other countries.

ISBN 978–1–137–45625–0

This book is printed on paper suitable for recycling and made from fully
managed and sustained forest sources. Logging, pulping and manufacturing
processes are expected to conform to the environmental regulations of the
country of origin.

A catalogue record for this book is available from the British Library.

A catalog record for this book is available from the Library of Congress.

Contents

Tables and Figures

Tables

Figures

Preface

This book has been produced to celebrate the 40th anniversary of the European Group (EG) for the Study of Deviance and Social Control. I have participated in 20 of the its conferences in many interesting places in Europe, often in conflict zones due to the group's tradition of solidarity and activism. I have always found the conference an inspiring and welcoming forum which has contributed greatly in shaping me as a criminologist.

I carry with me warm and sometimes funny memories from these events, from people who have become my friends, and from some who are no longer with us, such as Louk Hulsman, who would always greet us all with his warm smile and special laughter, and from Stan Cohen, to whom this book is a tribute, and whom I first met at my very first EG conference, in Prague in 1993.

My hope is that green criminology will have its place in the conferences to come, and that the group will continue to contribute to expanding the field.

I dedicate this volume to Vilde and to the children and future children of the contributors to this book, in the hope that during their lifetime this world will be a better place, for all.

Contributors

Andrea Beckmann is a critical criminologist and a social pedagogue. She has published widely on areas of 'body' politics, not least on 'sexualities', and her work on consensual sadomasochism has attracted international recognition. Her book *The Social Construction of Sexuality and Perversion* (2009) was nominated for the Bonnie and Vern L. Bullough Book Award, which is given to outstanding work in the field of sexology. She has also engaged in various academic publications that challenge Western neoliberal education systems and their underlying assumptions. Her work with Charlie Cooper has been subsumed into *The Oxford Handbook of Criminology* (2007) as part of cultural criminology. She has for many years been the English representative of the European Group for the Study of Deviance and Social Control.

Avi Brisman is Assistant Professor in the School of Justice Studies at Eastern Kentucky University in Richmond, Kentucky, USA. His writing has appeared in *Contemporary Justice Review*, *Crime, Law and Social Change*, *Crime Media Culture*, *Critical Criminology*, *International Journal for Crime, Justice and Social Democracy*, *Journal of Contemporary Criminal Justice*, *Journal of Criminal Justice and Security*, *Journal of Qualitative Criminal Justice and Criminology*, *Journal of Theoretical and Philosophical Criminology*, *Race and Justice*, *Social Justice*, *Theoretical Criminology* and *Western Criminology Review*, among other journals. He is the co-editor, with Nigel South, of *Routledge International Handbook of Green Criminology* (2013), co-editor, with Nigel South and Rob White, of *Environmental Crime and Social Conflict: Contemporary and Emerging Issues* (2015) and co-author, with Nigel South, of *Green Cultural Criminology: Constructions of Environmental Harm, Consumerism, and Resistance to Ecocide* (2014).

Samantha Fletcher is Lecturer in Sociology, Crime and Deviance at Staffordshire University, UK. Her research involves critical explorations of the changing forms of both coercive state power and manifestations of state violence in the context responses to forms of contestation and protest movements.

David Rodríguez Goyes is a PhD student at the Department of Criminology and Sociology of Law, University of Oslo, Norway. Born in Bogotá, Colombia, he holds a law degree (2010) and an MA in law, sociology and criminal policy (2013), both from the National University of Colombia. He is the co-author of *Determinantes Científicas Económicas y Socioambientales de la Bioprospección en Colombia* (2013). His research interests within the field of green criminology are biopiracy/bioprospecting and the environmental impacts of the biomedical industry's research practices. He holds a position as Lecturer in Criminology at the Antonio Nariño University of Colombia, where he is currently on leave to take a PhD in criminology at the University of Oslo.

Allison Gray is a PhD student at the University of Windsor, Ontario, Canada. During her master's, she was introduced to and developed research interests in the areas of green criminology and food crime alongside Dr Ron Hinch. Her passions have led to projects that concern food safety, the individualization of food activism and non-human animal experiences of legal regulation.

Joanna F. Hill is a doctoral student at University College London's (UCL's) Security Science Doctoral Research Training Centre, UK. She has a BSc in ecology and conservation from Anglia Ruskin University, Cambridge, and an MRes in biosystematics from Imperial College London, UK. She manages the UK National Environmental Crime Conferences, is a co-founder of UCL's Agent-Based Modelling Working Group and has volunteered for several UK and international conservation organizations. Her research interests focus on community participation in conservation and using agent-based simulation models to understand and prevent wildlife crime.

Ron Hinch is Professor in the Faculty of Social Science and Humanities at the University of Ontario Institute of Technology, Canada. His eclectic research interests include food crime (all the nasty things that are done to our food before we buy it), green criminology, critical criminology, criminological theory and serial homicide.

Katrin Kremmel is a junior researcher at the Institute for the Sociology of Law and Criminology in Vienna, Austria. Her main tasks lie within the FP7 project ALTERNATIVE, focusing on restorative practices in intercultural settings. She studied cultural and social anthropology with a central interest in legal anthropology and restorative justice.

Michael A. Long is Senior Lecturer in Social Sciences at Northumbria University, UK. His research focuses on green criminology, political economy and state–corporate crime. Some of his recent articles have appeared in *British Journal of Criminology*, *Competition & Change* and *Social Psychology Quarterly*. He recently co-authored, with Paul Stretesky and Michael Lynch, *The Treadmill of Crime: Political Economy and Green Criminology* (2013).

Michael J. Lynch is Professor of Criminology and an associate faculty member of the Patel School of Global Sustainability at the University of South Florida, USA. He has been engaged in green criminological research since 1990. His other areas of interest include radical criminology, environmental justice, racial biases in criminal justice practices, and corporate crime and its control.

Wayne Morrison is Professor of Law, Queen Mary University of London, UK, and barrister and solicitor of the High Court of New Zealand. He has travelled and worked globally with previous posts in Japan and stints in wine, food and sport. He was Director of the University of London's external programmes for law from 1999 to 2009. His research and publications span criminological and legal theory. Recently, he has worked towards a more global criminology that includes topics that are traditionally excluded from the canon, such as genocide, and methods of representation and remembrance. His most recent book on criminology is *Criminology, Civilisation and the New World Order* (2006, Spanish edition 2012).

Lorenzo Natali holds a PhD in criminal law and criminology and is currently a research fellow in the Department of Criminology at the University of Milano–Bicocca, Milan, Italy. His main fields of research so far have focused on violent crime, environmental crime, symbolic and radical interactionism, green criminology, cultural criminology, and qualitative and interdisciplinary approaches to green crime, including visual green criminology.

Brunilda Pali is currently a researcher and PhD candidate in the Leuven Institute of Criminology at the University of Leuven, Belgium, as part of the research team for the FP7 project ALTERNATIVE, working on restorative justice application in intercultural settings. She has a background in feminist and cultural studies, and her main interests relate to critical criminology, social justice, critical social theory and psychoanalysis.

Vincenzo Ruggiero is Professor of Sociology and Director of the Crime and Conflict Research Centre at Middlesex University, London, UK. He has conducted research on behalf of national and international agencies, including the Economic and Social Research Council, the European Commission and the United Nations. He has published in many international academic journals, and his books include *Crime and Markets* (2000), *Movements in the City* (2001), *Crime in Literature* (2003), *Understanding Political Violence* (2006), *Social Movements: A Reader* (2008), *Penal Abolitionism* (2010), *Corruption and Organised Crime in Europe* (2012), *The Crimes of the Economy* (2013) and *Power and Crime* (2015).

Ragnhild Aslaug Sollund is Professor of Criminology in the Department of Criminology and Sociology of Law at the University of Oslo, Norway. She is currently doing research on the legal and illegal trade in endangered animal species. Her research is also part of the EFFACE project entitled 'European Action to Fight Environmental Crime', financed by the European Union. While for many years doing research on migrant women (e.g. her dr. polit. degree), violence and police racial profiling, she has over the last decade contributed to many green criminology anthologies such as *Routledge International Handbook of Green Criminology* (2013, eds South and Brisman) and *Emerging Issues in Green Criminology* (2013, eds Walters, Westerhuis and Wyatt). She is the author of three monographs, sole editor and co-editor of several books, including *Global Harms: Ecological Crime and Speciesism* (2008), *Eco-global Crimes: Contemporary Problems and Future Challenges* (2012, eds Ellefsen, Sollund and Larsen) and *Critical Views on Crime, Policy and Social Control* (2014, eds Sorvatzioti, Antonopolous, Papanicoulaou and Sollund).

Nigel South is a professor in the Department of Sociology and the Centre for Criminology and a member of the Human Rights Centre and the Essex Sustainability Institute at the University of Essex, UK. He is also an adjunct professor in the School of Justice at Queensland University of Technology, Brisbane, Australia. In 2013, he received a Lifetime Achievement Award from the American Society of Criminology, Division on Critical Criminology, and he currently serves as European Editor for the journal *Critical Criminology*. With Avi Brisman, he is both co-editor of *Routledge International Handbook of Green Criminology* (2013) and co-author of *Green Cultural Criminology* (2014).

Paul B. Stretesky is Professor of Criminology in Social Sciences and Modern Languages at Northumbria University Newcastle, UK. His books include *Exploring Green Criminology* (2014, with M.J. Lynch), *Treadmill of Crime: Political Economy and Green Criminology* (2013, with M. Long and M.J. Lynch) and *Guns, Violence and Criminal Behavior: The Offender's Perspective* (2009, with M.R. Pogrebin and N.P. Unnithan). His book *Environmental Crime, Law and Justice* (2014, with M.J. Lynch and R. Burns) was recently released in its second edition. He is the co-editor (with M.J. Lynch) of the Ashgate book series *Green Criminology*.

1

Introduction: Critical Green Criminology – An Agenda for Change

Ragnhild Aslaug Sollund

The role of critical green criminology

This book is the outcome of the conference of the European Group (EG) for the Study of Deviance and Social Control, which was hosted by the Department of Criminology and Sociology of Law at the University of Oslo, Norway, in September 2013. The event marked the 40th anniversary of this important criminology group, which for over four decades has influenced European and international criminology, and not the least what is usually referred to as 'critical criminology' – a field of criminology which arose as a criticism of the ways in which (mainstream, positivist North American) criminology was a vehicle and support to the judicial system and punishment apparatus. This is also part of the history of the Institute for Criminology and Sociology of Law in Oslo, which was first established as the Institute of Criminology and Penal Law 60 years ago.

The EG was funded and held its first conference in Italy in 1973 on the theme of 'Deviance and Social Control in Europe: Scope and Prospects for a Radical Criminology' (EG website).[1] Well-known critical criminologists, such as Stan Cohen, Vincenzo Ruggiero, Jock Young, Nils Christie, Louk Hulsman, Barbara Hudson and Thomas Mathiesen, were present at this conference and/or have been present at other EG conferences over the years for shorter or longer periods of time, and they have influenced and continue to influence critical criminology. The aim of the foundation of the group by Stan Cohen, Mario Simondi and Karl Shumann was not just to cover topics and hold debates which would otherwise be marginalized or ignored by mainstream, administrative and official criminology, but to establish a new network that could support and provide solidarity with emerging social movements (EG website). The foci of the group were, and remain today, on the construction of harm (e.g. Hillyard et al., 2004; Hillyard and Tombs, 2007), the construction of crime (e.g. Christie, 1994) and a commitment to social justice, to explore how deviance is socially constructed and controlled (e.g. Scraton, 2007), not the least in, through and beyond the judicial and prison systems (Hulsman, 1986; Mathiesen, 2014; Hallsworth, 2000).

The group has always been solidary with those who are oppressed and has maintained a strong focus on gender – for example, women as victims (Leander, 2006), offenders and prisoners (McMahon, 1999). Very often the group meeting at the conference's final day has included writing and agreeing on a resolution in someone's support, and often in conflict zones – for example, Crossmaglen in Northern Ireland or in Cyprus where conferences have been arranged.

The role of critical criminology

Political activism has been central to the history of the EG, and the link between political activism and critical criminology remains strong. This is also the case in the present book, which has been produced by contributors to the stream 'green criminology and political activism' at the EG 2013 conference in Oslo.

The divide between mainstream criminology and (critical) green criminology persists today and, according to Lynch and Stretesky (2014), it is rooted in the tendencies of mainstream criminology to (still) 'exclude a diverse range of topics relevant to studying harms and their consequences that ought otherwise to fit within the discipline of criminology if criminology were not so narrowly defined in the first place' (2014: 4).

The role of critical criminology, or radical criminology (Lynch and Stretesky, 2014), is not merely to describe and analyse the crimes and harms that are committed by the powerful, whether states (e.g. Cohen, 2001), corporations (Pearce and Tombs, 2009), penal systems (Christie, 2007; Hulsman, 1991; Mathiesen, 1967) or the capitalist system (Stretesky et al., 2013). The aim is to change the current world order by rejecting capitalism and consumerism as the leading values of our time, by rejecting the unfair distribution of wealth and power, by rejecting the criminalization of the powerless, and by rejecting the exploitation of those who cannot defend themselves and those whom the police and judicial systems fail to protect.

Green critical criminology

As shown by South (2014), green criminology arose as a field when researchers more or less simultaneously at different places on earth developed the same concern about environmental harms and crimes (see also Lynch and Stretesky, 2014: 3). Although Michael Lynch's (1990) 'The greening of criminology' is often referred to as a starting point for the field, many were not made aware of this article. Research on environmental crime had been going on for example in Norway, regarding pollution at sea (Christophersen and Johansen, 1991), and elsewhere corporate environmental crime (Pearce and Tombs, 2009), although this work was not clearly defined as belonging to any specific field of criminology apart from being critical criminology. But as South says,

> in an important sense, a green criminology is justified because it was inevitable and necessary. It reflected scientific interests and political challenges of the moment, carried forward the momentum of critical non-conformist criminology, and offered a point of contact and convergence.
>
> (2014: 8)

The umbrella under which green criminology topics are gathered is wide, and according to Beirne and South (2007), it includes 'the study of those harms against humanity, against the environment including space, and against nonhuman animals committed both by powerful institutions (e.g. governments, transnational corporations, military apparatuses) and also by ordinary people' (2007: xiii). Researching what is legally understood as crimes – breaches of law – has therefore been expanded to include studies of harm, and the concept of crime is conceptually changed to encompass even those harmful acts which are actually not crime, legally defined, but yet are as harmful as any act which is actually a breach of a law or regulation (Beirne and South, 2007; Sollund, 2008; Walters, 2010; White, 2007, 2013: 8; Stretesky et al., 2013; Lynch and Stretesky, 2014; Hillyard et al., 2004; South, 2008), whether as part of daily practices (Agnew, 2013) or even when instigated by organizations such as the International Monetary Fund and the World Bank (O'Brien, 2008). This marks a clear distinction with conventional criminology with its focus primarily on (only) those acts which are criminalized, crime prevention, offenders and punishment.

An important issue is thus also the area in between legal and illegal crimes – the justifications for criminalizing some acts and not others, for punishing some and not others, even when the acts/crimes that have been committed are qualitatively the same, and correspondingly in regarding some as 'justified' victims and not others (Sollund, 2012b). Given the impact of political activism on green criminology, an important part of the field is to show who the victims of environmental harms are, and to propose policy implications 'which would "work" legally speaking as well as being conceptually robust' (Hall, 2014: 99).

There is as such no divide between critical criminology in its original form in the way it was first set out – for example, in the birth of the EG – and green, eco-global criminology (White, 2011). Green criminology is a natural and necessary prolongation and expansion of critical criminology (Sollund, 2014; South, 2014) in a world that faces even greater problems now than it did in the 1970s as a consequence of the capitalist order, in which colonialism cannot be talked about in a past tense; it has merely changed its form. It is no longer Western states which 'own' the colonial states – they are in practice to a large degree often 'owned' by large corporations, and these corporations have no other purpose than to increase the wealth of their owners at the cost of the inhabitants of these previous colonies, and by harming indigenous peoples whether in Greenland (Myrup, 2012), Canada, Nigeria (Zalik, 2010), Colombia (Mol, 2013) or Ecuador (Sollund, 2012) – for example, by the dumping of toxic waste (Bisschop and Walle, 2013; White, 2008).

Green criminology cannot be claimed to be a unitary field as there are today degrees of how critical (radical) the writers appear to be, and the ways in which this reflect their research agendas. Further, the interests of criminologists who are writing within a critical tradition about green (e.g. Beirne and South, 2007), eco-global (White, 2011; Ellefsen et al., 2012) topics are as diverse as are criminologists.

White (2013) categorizes the different branches of green criminology as radical green criminology (e.g. Stretesky et al., 2013; Lynch and Stretesky, 2014), eco-global criminology (White, 2011, 2012), conservation criminology (Gibbs et al., 2010), environmental criminology (Wellsmith, 2010; Hill, Chapter 10), constructivist green criminology (Brisman, 2012) and speciesist criminology[2] (e.g. Beirne, 2009; White, 2013a: 23–24). And one could add historical criminology (Beirne, 1994, 2009; Ystehede, 2012) and green cultural criminology (Ferrell, 2013; Brisman and South, 2014), to which I will return.

One question arises: What might unite all of these branches of green; what do they have in common? The development into different fields of green criminology suggests that not all authors with their different fields of interest and research agendas will agree on perspectives, for example, in eco-justice and species justice (Halsey, 2004). A conventional (conservationist) criminologist could, for example, focus on illegal wildlife trafficking, the main problem being that animals are taken in violation of laws or regulations; thus the discussion would circulate around how to prevent this as a (and because it is a) crime rather than as a harm (Pires and Clark, 2012; Warchol et al., 2004). Such a perspective would be regarded as too narrow for a radical, anti-anthropocentric, criminologist who would refer to such acts as crimes either way and be equally concerned about those who are killed or abducted legally as those who are abducted and killed illegally (Rodriguez Goyes, Chapter 9; Sollund, 2011).

As South points out (1998, also see White, 2013b), green criminology provides no theory which can unify the field. This could by some be regarded as a deficiency. However, to believe that one grand theory could possibly unite all aspects of this rich field would probably undermine its content because it would necessarily have to cover contradicting interests and perspectives – for example, those which can be implicit in environmental rights versus species justice. The strength of green criminology is, according to Avi Brisman, its inclusion of a broad set of interpretations, explanations and predictions, and in this way the scope of green criminology is broader, its potential greater and its significance far more profound (Brisman, 2014: 24). As observed by Brisman, green criminologists do not come to the field without previous experience or theoretical baggage, and they have an array of theories to pick from in their analyses. One thing that many green criminologists further have in common, at least those who position themselves within critical green criminology, is an insistent awareness of exploitative relationships and injustice (Wyatt et al., 2014).

One of the advantages of critical green criminology is also its multidisciplinary approaches (e.g. Gibbs et al., 2010): while often leaning on the natural sciences for empirical evidence about environmental harms (and crimes) (e.g. Stretesky et al., 2013; Lynch and Stretesky, 2014), it also encompasses numerous perspectives from the social sciences as well as philosophy (e.g. Beirne, 1999), thus enriching the analyses (e.g. Sollund, 2008; Ellefsen et al., 2012).

Justice is strongly present in 'the green criminology ideology' if one can speak of such in a unitary way. As White states, 'The green criminology perspective (therefore) tends to begin with a strong sensitivity towards crimes of the powerful and to be infused with issues pertaining to power, justice, inequality and democracy' (2103: 22).

Green critical criminology: Contested field(s)?

Not all critical criminologists, such as Stan Cohen to whom this book is a tribute (in particular, see Brisman and South, Chapter 2; Morrison, Chapter 3), would agree that a critical criminologist should engage herself in environmental problems and animal abuse. Cohen (2001: 289), like many others, asked:[3] 'Why should we occupy ourselves with such issues – in criminology – when there are so many abuses of human rights?'

The answer is simple: we do not have to choose; we can do both. It is possible to write about, to criticize and to engage in the various sides of exploitation and abuse no matter whether the victims are women, men, non-humans or the environment itself, hence focusing on all aspects of justice – eco-, species and environmental or atmospheric (Walters, 2013) – while discussing openly in which ways these are contradicting, which should be given priority (if any) and why.

It is quite well established that the different forms of 'isms' – sexism, racism, speciesism and so on – are shaped by the same patterns of discrimination and prejudice (Donovan and Adams, 1996; Cazaux, 1999). Therefore studying one form of abuse sheds light on how to understand another, and those oppressed by such 'isms' have the same right to concern, and in being recognized as victims, whether women who are victims of violence and abuse (e.g. Leander, 2006), human environmental victims (Natali, Chapter 4) or animal victims (Agnew, 1998; Beirne, 1999; Sollund, 2012b, 2013).

A critical radical criminology requires a critical victimology. There is considerable conceptual overlap between green criminology, with its focus on harm against the environment and non-human animals, and victimology, through its foci on interrogating and critically understanding the perpetration and aftermath of crime more generally. For a radical critical criminology it is crucial with a critical and reflective approach to understanding conceptualizations of victims and victimhood, such as how these are contested, and culturally and historically specific (Fitzgerald, 2010: 132,134).

Despite Cohen's apparent resistance to including non-human animal victims among those who deserve to be an object of study, thus including the mechanisms which cause them victimization – as much as the mechanisms which cause human victimization – his theories are at least as applicable within the field of critical green criminology as they are in more 'conventional' critical criminology (Brisman and South, Chapter 2). These include particularly his theories about denial

(Cohen, 2001) – for example, with regard to analyses of responses to climate change,[4] and harm more generally, including animal harm.

Concerning environmental crime/harm, I think it is easier to gain support for this to be a legitimate part of criminology today – and to be accepted as criminology, even when it is not related to the breach of a law or regulation – than when green criminology was in its infancy. The question has been raised: Is this criminology (e.g. O'Brien, 2008)? I think this question no longer needs to be asked, as observed by Walters et al. (2013: 12). Of course this is criminology, and the myriad of green criminology books which have been published over the past decade, plus the presence of green criminology panels at criminology conferences, including a presidential panel at the American Society of Criminology (ASC)[5] conference in 2013, clearly document the strength of this expanding field.

Animal abuse as part of green criminology: More contested?

Animal abuse, whether legal or illegal, has become an important part of green criminology (e.g. Beirne, 1999, 2007, 2008, 2011, 2013; Wyatt et al., 2014; Cazaux, 1998, 1999; Sollund, 2008, 2011, 2012, 2012a, 2012b, 2013; Nurse, 2013). Whether animal abuse should be included in criminology may be more contested, and it is also debated which parts of this specific field of green criminology rightfully belong there (Stretesky et al., 2013). Stretesky et al. are critical of the ways in which green criminologists include animal harm in the field and argue that only those forms of harm and abuse which can explicitly be connected to the treadmill of production can legitimately be placed within the field. They claim that factory farming is one such topic which should not be included in green criminology. This crime is, however, one which is very hard to disconnect from corporate crime/harm and capitalism. Think, for example, of McDonald's and the ways in which its[6] (and other corporations') production of cheap meat entails deforestation (Boekhout van Solinge and Kuijpers, 2013; Agnew, 2013; Mies and Shiva 2014) and massive animal abuse (Croall, 2013; Sollund, 2008).

When Stretesky et al. state that abuse of animals in corporate farms should be excluded from green criminology because 'they do not appear to involve any kinds of harm to ecosystems which is compatible to green criminology' (2013: 107), they are wrong:

> the true costs of industrial agriculture, and specifically 'cheap meat', have become more and more evident. Today, 'the livestock sector

emerges as one of the top two or three most significant contributors to the most serious environmental problems'... This includes stresses such as deforestation, desertification, 'excretion of polluting nutrients, overuse of freshwater, inefficient use of energy, diverting food for use as feed and emission of GHGs...'.

> (UNEP, 2012; see also Beirne, 2011: 354; Gray and Hinch, Chapter 6)

Corporate farming not only implies severe animal abuse and premature death for billions of animals every year, it is also a green crime with detrimental effects to ecosystems and in contributing to global warming because of the massive carbon emissions that are produced. The overall tendency is also an increase in industrial farming, with corporate farms with more animals, and consequently more abuse with, for example, the use of concrete, barren crates where the animals can neither stand nor turn, rather than small farms where the animals have a chance to be seen and considered as social individuals with needs, rather than production units (Sollund, 2008).

Corporate farming is, in my view, not only an acceptable topic for green critical criminology but an ideal one because of its multifaceted array of harms, and the ways in which these are empirically constructed and ideologically denied (Cohen, 2001) and neutralized (Sykes and Matza, 1957; Sollund, 2012b) by perpetrators, strong interest groups such as corporate agribusiness, politicians and consumers. It includes massive exploitation, abuse, theriocide (Beirne, 2014) and environmental destruction, such as deforestation (Agnew, 2013; Boekhout van Solinge and Kuijpers, 2013). What we eat is heavily informed by cultural practices and beliefs, what is accessible to us and how this is presented to us – for example, through state policy and advertisements – thus also making it relevant from a green, cultural criminology perspective (Brisman and South, 2014).

This is an example of why it is important that while the field of green, critical criminology is still fairly young and evidently expanding, no one should set out to 'police' the borders of the field in order to include or exclude particular research areas, and determine what kind of research and topics should be 'permitted' and which should not. How the different scholars choose to position themselves in the field is, I think, up to them. This said, I welcome debate because, most importantly, green, critical criminology is and should be open-minded and should include fields which hitherto have not been considered as 'criminology', and debate is a condition for theoretical refinement.

Inevitably, as shown above and by White on several occasions (2013a, 2013b), green criminology can encompass many fields and perspectives, which I regard as an enrichment rather than a problem.

A recent branch of green criminology is what Brisman and South (2014) call 'green cultural criminology'. They state that 'green criminology reflects a critical stance on the need to defend the environment and to uphold the rights and safety for both human and nonhuman species' (p. 5). They further argue that 'green criminology must attend to: the mediated and political dynamics surrounding the presentation of various environmental phenomena (especially news about environmental crimes, harms and disasters); the commodification and marketing of nature and examples of resistance to environmental harm and demand for changes in the way that "business as usual" and "ordinary acts of everyday life" (Agnew, 2013) are destroying the environment' (Brisman and South, 2014: 6). With reference to critical cultural criminologists such as Jeff Ferrel, Keith Hayward and Jock Young, they see cultural criminology as exploring the convergence of cultural and criminal processes in contemporary, social life (2014: 7).

That green criminology and cultural criminology are closely related is argued by Jeff Ferrell:

> By the nature of their subject matter, both green criminology and cultural criminology push against the conventional borders of criminology, and so tend to upset the definitional and epistemic order of the discipline. Likewise, both are open to exploring a range of social harms and social consequences whether these harms are conventionally defined as 'criminal'. Currently left outside the orbit of law and criminality, or even themselves propagated by the criminalization process.
>
> (Ferrell, 2013: 349)

Here should also be mentioned the beautiful cultural historical non-speciesist criminological analyses of artworks by Hogarth (Beirne, 2013) and Potter (Beirne and Janssen, 2014), and the ways in which animals are portrayed in these works, and thereby the role of animals in society at the time when this art were produced.

Green criminology thus continues to expand, incorporate and recognize the perspectives that are included in the different 'forms' of criminology, whether called merely critical, radical (e.g. Lynch and Stretesky, 2014) or cultural (Ferrell et al., 2008).

The agenda for a future green critical criminology

The increase in wealth in the North, conflicts stirred through resource drainage and exploitation (Brisman and South, 2014) and social conflicts and wars (the past few years in Syria and with the expansion of the Islamic State) will continue to increase the flow of refugee migration to the North, and with that xenophobia, refugees' deaths at sea, othering and social divides in the refugees' destination countries, in addition to the criminalization of immigrants (Sollund, 2012; Aas, 2013). Signs of such tendencies are, for example, Switzerland's 2013 referendum against immigration, which entails quotas for the number of immigrants allowed into the country, and the recent elections in Sweden where the right-wing Sverigedemokraterna (Swedish democrats) party received one-third of the votes. As global warming progresses, more people will be forced to migrate and compete for the same resources, while at the same time in the Western world the problem is not scarce resources but overconsumption, waste and pollution – a problem that is often exported to the poorer hemisphere (e.g. White, 2008; Bisschop, 2013) but is also visible in the West as the remaining pollution from pre-regulation times (Dybing, 2012). Regulation also invites new forms of crimes that demand attention, as in the case of carbon trading (Walters and Martin, 2014).

The role of green critical criminology is to counteract the harms that are caused by the agents working within and through power structures, to critically scrutinize, unmask and deconstruct them to show who the real perpetrators are, what possibilities they are provided with to perform their harms within a legal framework – for example, biopiracy through the patenting of seeds by large corporations (South, 2007; Gray and Hinch, Chapter 6; Goyes and South, 2014) the extraction of oil and gas (White, 2013), palm oil production (Mol, 2013), legal and illegal wildlife trafficking (Sollund, 2011; Wyatt, 2013) 'green- and welfare-washing' (Aaltola, 2012; Brisman, 2009) – and to explore how injustice is produced, whether ecological, species or environmental (Beirne, 1999, 2007, 2014; Beirne and South, 2007; White, 2013). Harms include:

> transgressions that are *harmful to humans, and nonhuman animals,* regardless of legality per se, environmental-related harms that are facilitated *by the state,* as well as *corporations and other powerful actors* insofar as these institutions have the capacity to shape official definitions of environmental crime in ways that allow, condone, or excuse environmentally harmful practices.
>
> (White, 2013: 20–21)

With the risk of sounding polemic and programmatic, I argue that we need not only a green, radical critical criminology but also a revolutionary criminology to lead us on the path towards a more just world, for all non-humans and humans – children, women and men. This would be a path that acknowledges justice not only on behalf of those who are here now but for those who will come after us (White, 2011), and to ensure that new individuals of the now existing species may continue to be born, rather than go extinct. In this world, material economic value must be downplayed and stop guiding our policies, actions and goals, and instead be exchanged for recognition of what truly brings the world a step forward: respect for each other's intrinsic value, whether human, non-human or nature itself (Halsey and White, 1998; Benton, 1998). This entails an appreciation of 'feminine values' – care and empathy, which guide our actions towards others (Donovan and Adams, 1996). When witnessing abuse, we should not turn away to avoid the discomfort that it causes us to watch the suffering of others and thereby silently accept it; we should not deny it (Cohen, 2001); we should speak. Being 'soft' is not a shame; it allows you to see beyond yourself and include the rest of the world in your social sphere. We see clearly where anthro-androcentrism (human-male centredness) has led us and what is applauded: capitalism and people of fortune are given the status of demi-gods in Western society, but they have nothing to be proud of, their fortunes are not based on virtues and they are tinted with shame. Whether talking about states' fortunes based on former colonialism (Galeano, 2010) or the current exploitation of oil and gas (as in Norway), transnational corporations' successes are based on the exploitation of nature and the domination of others – be it women, children, non-human animals or nature (e.g. in mining and in the patenting of seeds) – so from a green criminology perspective these fortunes are made through crimes. Those behind ruthless exploitation should be dethroned, and the victims of injustice should be acknowledged, their voices heard and their interests recognized, even though they themselves cannot speak for their rights.

Rather than keeping consumerism as our moral guide and obligation, and seeing humans as being detached from the rest of the natural world, we should acknowledge that we are all interdependent – that by ruining the earth, we also ruin ourselves. This is why in Ecuador and Bolivia, Mother Nature (Pachamama) is granted rights (Zaffaroni, 2013), and why, in Costa Rica, hunting is forbidden.[7]

Green criminology should dare to pose the questions that others avoid – for example, those relating to the problems of human population growth. As the population expands and with it agriculture and intensive farming, other species lose their habitats and biodiversity is diminished (Tilman et al., 2001; Boekhout van Solinge and Kuijpers, 2013). Animals are forced to live in close contact with humans and vice versa, resulting in human–animal conflicts (Woodroffe et al., 2005), in which the animals are doomed to lose. Lynch et al. (Chapter 7) argue that green criminology should focus on the overarching structures in society which cause, for example, species extinction, something which I support. However, in this process it is important not to lose sight of the billions of individuals who are suffering whether as a result of social disorganization, capitalism, trafficking or loss of habitat, just as we would not avoid addressing a crime such as homicide simply because it is (perhaps) not structurally determined and does not entail a threat to the human species. Green criminology must therefore maintain its focus on the varied causes of crimes, whether we refer to a crime in a legal sense or not, and whether it affects one or many. Often a perspective that includes structural factors provides an important background to understanding social, cultural and individual practices which harm one individual, many and also entire species, but such harms must also be analysed through case studies and ethnographic research, to gain in-depth insights into why people participate in practices which may be counter to their ethics and moral standards. This is the role of a radical, green criminology as I see it.

In the spirit of the EG with its support of activists, critical criminology is concerned about the opportunities that people have to protest against harmful activities that are generated by powerful capitalist interests. The answer from the authorities when people protest against harms and injustice is often criminalization and persecution, whether of individual farmers protesting against international corporations which deprive them of their opportunity to farm in the way that they have traditionally (Mies and Shiva, 2014; White, 2011; Mol, 2013; Gray and Hinch, Chapter 6; Goyes and South, 2014), environmental movements protesting against oil extraction and oil from tar sand production (Zalik, 2010, 2012), animal rights activists or freedom-of-speech activists, such as the persecution of Edward Snowden and Wikileaks, whether they are protesting against exploitation, abuse or silencing (Aaltola, 2012; Ellefsen, 2012; Kremmel and Pali, Chapter 13).

Critical criminology, whether defined as green, cultural or eco-global, with its roots for example in the classic study by William Foot Whyte (1943), has long traditions for including bottom-up perspectives and ethnographic work in relation to prison research (Mathiesen, 1967), or more recently regarding protest movements (Ellefsen, 2012; Aaltola,

2012). These approaches, which combine critical views of power structures, of inequalities and how protest is responded to, are very much present in this volume.

Presentation of chapters

Chapter 2, 'State–corporate environmental harms and paradoxical interventions: Thoughts in honour of Stanley Cohen' by Avi Brisman and Nigel South, is a tribute to Cohen through which his work is applied to green criminology, thereby showing the relevance of his ideas in this field. Although he did not apply his ideas to green criminology himself, Cohen was concerned with the crimes and harms perpetrated by legitimate bodies, such as states and corporations, and with the way in which these acts and omissions, their impacts and consequences, have routinely been ignored, overlooked or denied. The chapter considers some examples of such crimes and harms from across recent decades, focusing on environmental degradation and destruction, and taking as a template the spirit and insights of Cohen's work. The authors cite Cohen (2001: 289), who admits to being 'utterly unmoved' by environmental and animal rights issues. Still, through the examination of how environmental destruction is the consequence of tactics or tools of war – for example, in Afghanistan, the Iraq War and in the Vietnam conflict – the applicability of Cohen's theories is shown. A further exploration of his theories is presented through another case – the exportation and diffusion of new biotechnologies to developing nations in more recent decades – which had as a consequence the intensification of production, with small farmers being squeezed out and inequality being increased, in addition to the threat to biodiversity through the promotion of monocultures. The chapter concludes by paralleling ecocide to genocide – which through his dedication to human rights was Cohen's concern – underlining the universality of such crimes, and in the hope that the adoption of an international crime of ecocide to sit alongside the crime of genocide would have been supported by Cohen.

Chapter 3, 'Looking into the abyss: Bangladesh, critical criminology and globalization' by Wayne Morrison, continues the homage to Stan Cohen, and to Jock Young and Stuart Hall, to discuss the war crimes processes in Bangladesh that resulted in verdicts in 2013, which were met by demonstrations, both for and against. The chapter starts with a theoretical discussion of what it means to 'be a critical criminologist', applying Nietzsche's anti-reason claim question: 'is seeing itself – not seeing abysses?'. It discusses the original meaning of 'theory', which meant to look at, to see, to contemplate through classic narratives of theoria which said 'that one need[s] to be taken [away] from one's habitual surroundings to see afresh'. Morrison was definitely taken out of his familiar surroundings when he first visited Bangladesh. He applies these experiences in a discussion of how the critical criminology classics can be used to throw theoretical light on the events in Bangladesh that are related to the crime processes and the conditions of Bangladesh garment workers. Expressed through the concept of vertigo (Young, 2007), the inhabitants of the Western world may feel a sense of emptiness amid their consumption. The object of consumption may seem to be a 'fake', but for the garment workers their hollowness is caused by physical hunger and thirst: 'There, in the areas surrounding Dhaka, the object-to-be-consumed is all-too-real, the stitching of which has to be perfect else penalties may befall, wages withheld.' Cohen (2001) was cited by the Second Tribunal of the War Crimes Tribunals of Bangladesh – 'after generations of denials, lies, cover-ups and evasions, there is a powerful, almost obsessive, desire to know exactly what happened' – to align the court with a recognizable ethical imperative. Lastly, Morrison carries out a reading of *Policing the Crisis* by Stuart Hall et al. (2013) to analyse the war processes, before finally arguing that the foundation of the everyday is the abyss: 'Under globalisation criminology must confront that'.

Chapter 4, 'A critical gaze on environmental victimization' by Lorenzo Natali, starts with a theoretical examination of critical and green criminology literature, showing the difficulties in establishing a clear link between cause and effect regarding environmental victimization, and how denial techniques further make this difficult. It further examines processes of environmental victimization in order to analyse the various steps, and he emphasizes that it is necessary to take into consideration the perception of harm 'from the inside', 'starting from the symbolic and cultural perspectives expressed by the social actors affected', and more precisely asking how people who are affected by pollution live and make meaning of their experiences in polluted places. Victims do not always recognize themselves as victims, and so, Natali claims, focusing on the mechanisms of denial will also help one to understand silence, apathy and a range of other possible responses by those who daily witness the destruction of their environment. This chapter thus also provides an interesting use of Cohen. In the final section, and echoing much of the agenda of critical criminology, Natali suggests a processual analysis of environmental harm, through which an environmental problem may turn into radical action and social change. Because powerful actors can use their influence to delay and/or weaken the efficacy of the law, it is crucial to facilitate the various steps towards social and political recognition and visualization of any harm that is connected to industrial activity. Finally comes a suggestion for a re-evaluation of the concept of 'fraternity, to guide our behaviour', to transform it towards our environment, making it ecologically and socially tenable, through a gradual development of an ethical protection of the environment and of all of those within it.

Chapter 5, ' "Creative destruction" and the economy of waste' by Vincenzo Ruggiero, is an examination of the relationship between the official economy and organized crime. The first section focuses on the illegal dumping of waste. In describing environmental harm, it shows that this is caused by both illegitimate and legitimate behaviour. After defining the legal framework for environmental harm, it draws on the categories that were utilized by Max Weber in his study of the relationship between law and economy. Part of the discussion concerns the normative effect of the law in perspective of daily practices. Through his reading of Weber, Ruggiero states: 'In brief, looking at environmental crime, we are faced with economic subjects engaged in instrumentally rational action, determined by the calculated end of profit; value-rational action, determined by the belief that bending rules is an ethical necessity; affectual action, inspired by the emotional aspect of material gain; and in traditional action, that is, "determined by ingrained habituation".' A discussion of the variables "innovation" and "deviance" – concepts which are used in both economics and criminology/sociology (Merton, 1968) – ends the chapter, criticizing the logic of economic growth. Ruggiero argues 'that growth is criminogenic not only because it can be metaphorically equated to obesity, but also because it depicts greed and acquisitiveness in a positive light, making them core values of individual and collective behavior'. Further he concludes that economic growth 'exacerbates polarization of wealth, and increases relative deprivation, one of the central variables in the analysis of crime . . . A radical critique of economic growth, therefore, could be a first step towards the prevention of environmental crime'.

Chapter 6, 'Agribusiness, governments and food crime: A critical perspective' by Allison Gray and Ron Hinch, also pursues 'legal crimes' in the analysis of the relationship between humans and the agricultural environment. Practices involving industrialization, corporatization and neoliberalization have drastically reformed the modern methods of the food industry. The chapter explores four specific cases: slave labour in the cocoa industry, including a historical presentation of the case; the impact of agribusiness on farming, including a description and discussion of its environmentally damaging effects, such as those caused by animal waste pollution (which comes in addition to the harm against and exploitation of the animals involved); the use of pesticides in agriculture, showing the history of the ways in which these are used and contested and how they damage human, animal and environmental health; and the impact of genetically modified foods on farming autonomy, including a discussion of the ways in which small-scale farmers are forced by agribusiness giants such as Monsanto to quit ancient practices of retaining a percentage of each year's crop to plant the next season, and when pursuing old practices they are confronted by lawsuits from Monsanto that they must lose owing to a lack of resources. These cases exemplify the ways in which current laws, where they exist, are ineffective in providing safe environments, nourishing food and healthy people, while being (re)constructed, neglected and overridden by agribusiness, in collaboration with governing bodies both locally and globally. They are eloquent examples of the crimes of the powerful, and the ways in which governing bodies and agrotech industries have consistently worked together over time to promote the interests of the latter through laws and regulations.

Chapter 7 by Michael J. Lynch, Michael A. Long and Paul B. Stretesky, 'Anthropogenic development drives species to be endangered: Capitalism and the decline of species', aims to fill what the authors see as a gap in the green criminology literature: the hitherto absence of a systemic analysis of structural factors which threaten non-human species. The reason for this focus is that, while case studies of wildlife trade and hunting have revealed important insights into such practices and their consequences, the major dangers to non-human species are caused by human development more generally, which has widespread impacts on species by destroying ecosystems and, on a larger scale, impeding ecosystem functionality and habitat structures through processes such as ecosystem segmentation. By 'systemic' is understood those ecological harms that are endemic to capitalism as a system of production – more specifically the forms of ecological disorganization that are produced by capitalism in its ordinary course of development – and which in the current context of global capitalism are global in their appearance, and therefore are structural in origin. The empirical data used to analyse the anthropocene, capitalist consequences for other species are reviews of the International Union for Conservation of Nature's Red List of threatened species in the USA and additional data on species endangerment from the US Wildlife and Fish Service. The data illustrate the extent of the harm that structural factors may cause to non-human animals. The authors conclude that future work on species decline must focus on structural factors, rather than case studies of different forms of threats to animal species, and thus not fail to get to grips with the broader economic forces that drive species harm.

Chapter 8, 'The illegal wildlife trade from a Norwegian outlook: Tendencies in practices and law enforcement' by Ragnhild Sollund, is perhaps exactly the kind of study which is criticized by Lynch et al. (Chapter 7) as being inadequate as a result of its case study-approach. The intention is to see what practices concerning legal and illegal wildlife trafficking are present in Norway, Colombia and Brazil through an examination of empirical data that were collected for an ongoing research project. The data include confiscation reports from customs departments and interview data with experts – representatives of bodies in charge of wildlife in Norway, Colombia and Brazil; individual stakeholders; and experts and with law-enforcement agencies. The chapter starts by presenting international trends in the legal and illegal wildlife trade in terms of the species used and for what purpose. It proceeds to show how these trends are represented in the data, mainly from the Norwegian study. The chapter includes a discussion of whether law enforcement of these crimes is adequate, efficient and a deterrent. One finding is that dealing with the illegal trafficking of CITES-listed[8] animals is poorly prioritized by police and customs, that the crime is usually regarded as a misdemeanour and that any punishment simply involves a minor fine. Animals in Norway are mostly killed when seized by customs or police. These findings are discussed from the perspective of green non-anthropocentric criminology.

Chapter 9, 'Denying the harms of animal abductions for biomedical research' by David Rodríguez Goyes, addresses the non-human suffering, the environmental harm and the detrimental effects for marginalized human communities that stem from experimental research into malaria. The aim is to analyse from a green criminology perspective the environmental harm production of the biomedical industry, illustrating that experimenting on animals is a source of systematic harm that is perpetrated against animals as well as against the environment and humans. The chapter starts by presenting theories of denial (Cohen, 2001) and neutralization techniques (Sykes and Matza), and then proceeds to looking at how the Patarroyo case in Colombia can be understood within this theoretical framework, and advancing the search for adequate guidelines to contextually determine when a situation deserves acknowledgment as being harmful and illegitimate. The

Patarroyo case involved the legal and illegal abduction of thousands of nocturnal monkeys from the Amazon basin to be used in malaria experiments by Dr Patarroyo. The chapter concurs with Sollund's (2012a) view that animal abuse often goes largely unquestioned, and it argues that this also applies to biomedical practices. The claim put forward is that scientific practices are backed by the modern idea of progress, and it is maintained that such ideology allows the denial of the myriad harms that are inflicted on non-human animals, natural environments and humans by this kind of research, and at the same time the labelling of its opponents as moralistic and blinded by environmental sentiments. The chapter concludes that the only way to alleviate the harms that are produced by the biomedical industry is to critically question the modern ideology of progress through the notion of species justice as a yardstick for acceptable practices. It enters a discussion about environmental justice and ecojustice vs. species justice in the contested field of animal experimentation.

Chapter 10, 'A systems thinking perspective on the motivations and mechanisms driving wildlife poaching' by Joanna F. Hill, explores the mechanisms that drive the illegal taking and killing of wildlife ('poaching'). It aims to define and explore the way in which poaching has been framed by instrumental and normative theoretical approaches. Drawing on ongoing doctoral research, Hill explains why a 'systems thinking' and computer simulation modelling approach can be used by both critical and empirical researchers to develop their understanding of environmental crime. Hill finds that rather than poachers belonging to a homogeneous group, people kill and take animals for a myriad of complex reasons that depend upon the immediate situation and a person's norms and values. To analyse these, various theoretical approaches are applied. Differential association theory suggests that when people interact with their peers they become socially conditioned with norms and values that justify their involvement with poaching, while also learning the practical methods to hunt. In much the same way, 'neutralization techniques' include the offenders' way of rationalizing, justifying and reducing the cognitive dissonance that they feel with regard to their deviant behaviour (Sykes and Matza, 1957). Hill argues why integrating instrumental and normative perspectives through the lens of systems thinking and computer simulation modelling has the potential to develop effective and ethical solutions to poaching especially, and environmental crimes more generally.

Chapter 11, ' "Now you see me, now you don't": About the selective permissiveness of synoptic exposure and its impact' by Andrea

Beckmann, critically explores the potential wider meanings and implications of a recent ruling of the European Court of Human Rights (2013) that reinforced a UK ban (2005) (*Animal Defenders International vs. the United Kingdom*) of an advertisement that was produced by an animal rights group about the legal trafficking of primates for the laboratory industry. This criminological exploration is informed by insights of critical theory/pedagogy and can be seen to represent critical, post-modern feminist as well as green-cultural criminological concerns and perspective(s). The ban was proclaimed to protect the public and to ensure a level playing field under Section 321 of the Communications Act 2003. Despite the claimed intention of the act, corporations do have considerably more access to financial and political modes of power than the public, and thus the reference to an 'even playing field' can only be considered to be a 'mystification of reality'. The criminalization of the animal rights activists and the ways in which this chapter addresses it have firm critical criminology roots, and can be seen in the light of other studies that focus on what appears to be a trend – the criminalizing of animal rights activists by powerful actors who gain from animal exploitation (Ellefsen, 2012; Aaltola, 2012).

Chapter 12, 'The Occupy movement vs. capitalist realism: Seeking extraordinary transformations in consciousness' by Samantha Fletcher, addresses another kind of activism and the experiences of the activists involved. As with the previous chapter, it shows that the state in being confronted often reacts with harsh responses. Using the Occupy movement as a vehicle for discussion, the chapter identifies and explores a series of state responses to contemporary protest movements. Drawing upon the first set of emerging published literature and narratives from the Occupy movement, alongside findings from preliminary empirical data that have been collated from qualitative interviews with activists from Occupy sites in the UK, the focus is on exploring the gamut of challenges that face contemporary activists today through a critical discussion of state responses towards Occupy, followed by some initial reflections on the potential significance of these responses. A Gramscian view of the state is employed and the chapter is divided into two themed sections that discuss state responses under the banners of 'coercion' and 'consent'. Topics for discussion in these sections include key themes in the policing of the Occupy movement and a review of the internal dynamics of the movement, both of which are used to problematize the ever-changing nature of state power. Fletcher concludes: 'Social movements, activism, protest and resistance remain key areas to turn our attention to, especially in terms of exploring the responses to them and those within them as a means to critique hierarchies of power, foster emancipatory knowledge and challenge existing power relations.'

Chapter 13, 'Refugee protests and political agency: Framing dissensus through precarity' by Katrin Kremmel and Brunilda Pali, is about yet another form of protest – that connected to the treatment of refugees in Austria. The authors state that during the last two decades we have witnessed an unprecedented number of refugee protests, which have profoundly challenged both the humanitarian and the security discourses about refugees. While the humanitarian discourse speaks of a subject in need of care and protection, the security discourse points at the refugee as a 'dangerous' agent who disturbs the public/national order, thus entailing the previously mentioned criminalizing of asylumseekers and migrants (Sollund, 2012; Aas, 2013). What both discourses have in common is the silencing of the political agency of the refugees. The authors read the refugee movements, focusing on the case of the Vienna protests, which started in 2012, by framing the refugees as political agents and their actions as constituting political action. They begin by contextualizing the debate about political agency through the work of Hannah Arendt on statelessness. This entails the loss of the 'right to have rights' and, according to Arendt, the political agency of a subject who lacks the 'right to have rights' is unthinkable. Within her framework of thought, refugees would have to belong to a body politic before they could make demands as political subjects. Furthermore, she believes, political action must be free of life necessities. Arguing with Jacques Rancière and Judith Butler, and applying their ideas of politics as dissensus and precarity to the reading of the refugee protest in Vienna, the authors challenge both of these claims, and they argue that the protest constitutes political action, precisely because the refugees take agency in spite of their lack of the 'right to have rights', through the staging of their precarity and life necessities.

A common feature of the last three chapters is that, through protesting, activists become a target for the authorities and a legitimate goal for oppression of meaning and action. In the last case, the fact that the activists in question are refugees adds to their vulnerability to attack by state authority that their power protests can engender. Themes which are repeated in this volume are the critique against the current system ruled by the capitalist order and the ways in which this jeopardizes environmental, eco- and species justice and rights. Through the ways in which this system is upheld by powerful agents, whether governments or transnational corporations, rights are breached and injustice is produced, and techniques of denial and neutralization of these harms are repeatedly applied.

Notes

1. http://www.europeangroup.org/content/history-values. Accessed on 30 September 2014.
2. The category of speciesist criminology is in my view unfortunately named because it indicates that those who adhere to this particular field are advocating speciesism rather than the opposite. It also belongs to radical criminology in being critical of anthropocentrism and in favour of species justice.
3. Personal communication.
4. Avi Brisman's presentation at the ASC conference in San Francisco in 2014 including a television clip about the imbalance in who are invited to debate global warming, through which climate deniers (the minority) are given as much room as the climate scientists, thus providing some interesting examples of denial.
5. https://www.asc41.com/Annual_Meeting/2013/Presidential%20Papers/ South,%20Nigel-White,%20Rob.pdf, https://www.asc41.com/Annual_Meeting/ 2013/Presidential%20Papers/Sollund%20Animal%20Abuse.pdf.
6. See, for example, http://www.mcdonaldscruelty.com/.
7. http://www.reuters.com/article/2012/12/11/us-costarica-hunting-idUSBRE8BA 04P20121211.
8. CITES: the Convention on International Trade in Endangered Species of Wild Fauna and Flora.

References

Aaltola, E. (2012) 'Differing philosophies: Criminalisation and the Stop Huntingdon Animal Cruelty debate'. In Ellefsen, R., Sollund, R. and Larsen, G. (eds) *Eco-global Crimes: Contemporary Problems and Future Challenges*. Farnham: Ashgate, pp. 157–181.

Aas, K.F. (2013) *Globalization and Crime*. London: Sage.

Agnew, R. (1998) 'The causes of animal abuse: A social-psychological analysis', *Theoretical Criminology*, 2(2): 177–209.

Agnew, R. (2013) 'The ordinary acts that contribute to ecocide: A criminological analysis'. In South, N. and Brisman, A. (eds) *The Routledge International Handbook of Green Criminology*. London: Routledge, pp. 58–73.

Beirne, P. (1994) 'The law is an ass: Reading EP Evans', *The Medieval Prosecution and Capital Punishment of Animals'*, *Society and Animals*, 2(1): 27–46.

Beirne, P. (1999) 'For a nonspeciesist criminology: Animal abuse as an object of study', *Criminology*, 37(1): 117–148.

Beirne, P. (2007) 'Animal rights, animal abuse and green criminology'. In Beirne, P. and South, N (eds) *Issues in Green Criminology*. Devon: Willan, pp. 55–83.

Beirne, P. (2009) *Confronting Animal Abuse: Law, Criminology, and Human-Animal Relationships*. Lanhan: Rowman and Littlefield.

Beirne, P. (2013) 'Hogarth's animals', *Journal of Animal Ethics*, 3(2): 133–162.

Beirne, P. (2014) 'Theriocide: Naming animal killing', *International Journal for Crime, Justice and Social Democracy*, 3(2): 50–67.

Beirne, P.S. and Janssen, J. (2014) 'Hunting worlds turned upside down: Paulus Potter's *Life of a Hunter*', *Tijdschrift over Cultuur & Criminaliteit*, 2(4): 15–28.

Beirne, P. and South, N. (eds) (2007) 'Introduction: Approaching green criminology'. In Beirne, P. and South, N. (eds) (2013) *Issues in Green Criminology*. Devon: Willan, pp. xiii–xxii.

Benton, T. (1998) 'Rights and justice on a shared planet: More rights or new relations?' *Theoretical Criminology*, 2(2): 149–175.

Bisschop, L. and Walle, G.V. (2013) 'Environmental victimisation and conflict resolution: A case study of e-waste'. In Westerhuis, D., Walters, R. and Wyatt, T. (eds) *Emerging Issues in Green Criminology: Exploring Power, Justice and Harm*. London: Palgrave Macmillan, pp. 34–57.

Boekhout van Solinge, T. and Kuijpers, K. (2013) 'The Amazon rainforest: A green criminological perspective'. In South, N. and Brisman, A. (eds), *Routledge International Handbook of Green Criminology*. London: Routledge, pp. 199–214.

Brisman, A. (2009) 'It takes green to be green: Environmental elitism, ritual displays, and conspicuous non-consumption', *Notre Dame Law Review*, 85: 329.

Brisman, A. (2012) 'The cultural silence of climate change contrarianism'. In White, R. (ed.) *Climate Change from a Criminological Perspective*. New York: Springer, pp. 41–70.

Brisman, A. (2014) 'On theory and meaning in green criminology', *International Journal for Crime, Justice and Democracy*, 3(2): 22–35.

Brisman, A. and South, N. (2014) *Green Cultural Criminology: Constructions of Environmental Harm, Consumerism, and Resistance to Ecocide*. London: Routledge.

Cazaux, G. (1998) 'Legitimating the entry of the animals issue into (critical) criminology', *Humanity & Society*, 22(4): 365–385.

Cazaux, G. (1999) 'Beauty and the beast: Animal abuse from a non-speciesist criminological perspective', *Crime, Law and Social Change*, 31(2): 105–125.

Christie, N. (1994) *Crime Control as Industry: Towards Gulag, Western Style*. London and New York: Routledge.

Christie, N. (2007) *Limits to Pain: The Role of Punishment in Penal Policy*, New York, Eugene, OR: Wipf & Stock Pub.

Cohen, S. (2001) *States of Denial: Knowing about Atrocities and Suffering*. Cambridge: Polity Press.

Croall, H. (2013) 'Food crime. A green criminology perspective'. In Beirne, P. and South, N. (eds) *Routledge International Handbook of Green Criminology*. London and New York, pp. 167–184.

Donovan, J. and Adams, C.J. (eds) (1996) *Beyond Animal Rights: A Feminist Care Tradition in Animal Ethics: A Reader*. Columbia: Columbia University Press.

Dybing, S. (2012) 'Environmental harm: Social causes and shifting legislative dynamics'. In Ellefsen, R., Larsen, G. and Sollund, R. (eds) *Eco-global Crimes: Contemporary Problems and Future Challenges*. Farnham: Ashgate, pp. 273–295.

Ellefsen, R. (2012) 'Green movements as threats to order and economy: Animal activists repressed in Austria and beyond'. In Ellefsen, R., Solund, R. and Larsen, R. (eds) *Eco-global Crimes: Contemporary Problems and Future Challenges*. Farnham: Ashgate, pp. 181–209.

Ellefsen, R., Larsen, G. and Sollund, R. (eds) (2012) *Eco-global Crimes: Contemporary Problems and Future Challenges*. Farnham: Ashgate.

Ferrell, J. (2013) 'Tangled up in green: Cultural criminology and green criminology'. In South, N. and Brisman, A. (eds) *Routledge International Handbook of Green Criminology*. London: Routledge, pp. 349–365.

Ferrell, J., Hayward, K. and Young, J. (2008) *Cultural Criminology: An Invitation*. London: Sage.

Fitzgerald, A.J. (2010) 'The "underdog" as "ideal victim"? The attribution of victimhood in the 2007 pet food recall', *International Review of Victimology*, 17(2): 131–157.

Galeano, E. (2010) *Las venas abiertas de America Latina*. Madrid: Siglo XXI de España Editores, S.A.

Gibbs, C., Gore, M.L., McGarrell, E.F. and Rivers, L. (2010) 'Introducing conservation criminology towards interdisciplinary scholarship on environmental crimes and risks', *British Journal of Criminology*, 50(1): 124–144.

Goyes, D.R. and South, N. (2014) 'The theft of land and nature: Land-grabs, biopiracy and the inversion of justice', paper presented at the ASC conference, San Francisco, 21 November 2014.

Hall, M. (2014) 'The role and the use of law in green criminology', *International Journal for Crime, Justice and Social Democracy*, 3(2): 97–110.

Hall, S., Critcher, C., Jefferson, T., Clarke, J. and Roberts, B. (2013) *Policing the Crisis: Mugging, the State and Law and Order*. Basingstoke: Palgrave Macmillan.

Hallsworth, S. (2000) 'Rethinking the punitive turn: Economies of excess and the criminology of the other', *Punishment & Society*, 2: 145.

Halsey, M. (2004) 'Against "green" criminology"', *British Journal of Criminology*, 44(6): 833–853.

Halsey, M. and White, R. (1998) 'Crime, ecophilosophy and environmental harm', *Theoretical Criminology*, 2: 345–371.

Hillyard, P., Pantazis, C., Tombs, S. and Gordon, D. (eds) (2004) *Beyond Criminology? Taking Harm Seriously*. London: Pluto Press.

Hillyard, P. and Tombs, S. (2007) 'From "crime" to social harm?' *Crime Law Social Change*, 48: 9–25.

Hulsman, L. (1991) 'Abolitionist case: Alternative crime policies', *The Israel Law Review*, 25: 681.

Hulsman, L.H. (1986) 'Critical criminology and the concept of crime', *Crime, Law and Social Change*, 10(1): 63–80.

Johansen, P.O. (ed.) and Christophersen, J.G. (1991) *Studier i økonomisk kriminalitet*. Oslo: Institutt for kriminologi og strafferett.

Leander, K. (2006) 'Reflections on Sweden's measures against men's violence against women', *Social Policy & Society*, 5(1): 115–125.

Lynch, M. (1990) 'The greening of criminology: A perspective on the 1990s', *The Critical Criminologist*, 2(3): 1–4.

Lynch, M.J. and Stretesky, P.B. (2014) *Exploring Green Criminology: Toward a Green Criminological Revolution*. Farnham: Ashgate.

McMahon, M. (1999) 'Assisting female offenders: Art or science', *Women and Girls in the Criminal Justice System: Policy Issues and Practice Strategies*. Kingston, NJ: Civic Research Institute, pp. 2–18.

Mathiesen, T. (1967/2012) *The Defences of the Weak (Routledge Revivals): A Sociological Study of a Norwegian Correctional Institution*. London: Routledge.

Mathiesen, T. (2014) *The Politics of Abolition Revisited*. London and New York: Routledge.

Merton, R. (1968) *Social Theory and Social Structure*. New York: The Free Press.
Mies, M. and Shiva, V. (2014) *Ecofeminism: Atlantic Highlands*. New Jersey: Zed Books.
Mol, H. (2013) 'A gift from the tropics to the world': Power, harm, and palm oil'. In Walters, R., Westerhuis, D.S. and Wyatt, T. (eds) *Emerging Issues in Green Criminology: Exploring Power, Justice and Harm*. Basingstoke: Palgrave, pp. 242–260.
Myrup, M. (2012) 'Industrialising Greenland: Government and transnational corporations versus civil society'. In Ellefsen, R., Larsen, G. and Sollund, R. (eds) *Eco-global Crimes: Contemporary Problems and Future Challenges*. Farnham: Ashgate, pp. 257–273.
Nurse, A. (2013) *Animal Harm: Perspectives on Why People Harm and Kill Animals*. Farnham: Ashgate Publishing.
O'Brien, M. (2008) 'Criminal degradation of consumer culture'. In Sollund, R. (ed.) *Global Harms: Ecological Crime and Speciesism*. New York: Nova Science Publishers, pp. 35–51.
Pires, S. and Clarke, R.V. (2012) 'Are parrots CRAVED? An analysis of parrot poaching in Mexico', *Journal of Research in Crime and Delinquency*, 49(1): 122–146.
Pearce, F. and Tombs, S. (2009) 'Toxic capitalism: Corporate crime and the chemical industry'. In Whyte, D. (ed.) *Crimes of the Powerful: A Reader*. Maidenhead: Open University Press, pp. 93–98.
Pires, S. and Clarke, R.V. (2012) 'Are parrots CRAVED? An analysis of parrot poaching in Mexico, *Journal of Research in Crime and Delinquency*, 49(1): 122–146.
Scraton, P. (2007) *Power, Conflict and Criminalisation*. New York: Routledge.
Sollund, R. (2008) 'Causes for speciesism: Difference, distance and denial'. In Sollund, R. (ed.) *Global Harms: Ecological Crime and Speciesism*. New York: Nova Science Publishers, pp. 109–131.
Sollund, R. (2011) 'Expressions of speciesism: The effects of keeping companion animals on animal abuse, animal trafficking and species decline', *Crime, Law and Social Change*, 55(5): 437–451.
Sollund, R. (2012a) 'Oil production, climate change and species decline: The case of Norway'. In White, R. (ed.) *Climate Change from a Criminological Perspective*. New York: Springer, pp. 135–147.
Sollund, R. (2012b) 'Speciesism as doxic practice versus valuing difference and plurality'. In Ellefsen, R., Sollund, R. and G. Larsen (eds) *Eco-global Crimes: Contemporary Problems and Future Challenges*. Farnham: Ashgate, pp. 91–115.
Sollund, R. (2012c) *Transnational Migration, Gender and Rights*, Vol. 10. Croydon: Emerald Group.
Sollund, R. (2013) 'Crimes understood through an ecofeminist perspective'. In South, N. and Brisman, A. (eds) *Routledge International Handbook of Green Criminology*, pp. 317–331.
Sollund, R. (2014) 'Øko-global kriminologi og feltets berettigelse eksemplifisert ved en empirisk studie'. In Finstad, L. and Lomell, H. M. (eds) *Motmæle: en antologi til Kjersti Ericsson, Cecilie Høigård og Guri Larsen* (red.) Oslo: Unipub, pp. 381–399.
South, N. (1998) 'A green field for criminology? A proposal for a perspective', *Theoretical Criminology*, 2(2): 211–233.

South, N. (2008) 'Nature, difference and the rejection of harm: Expanding the agenda for green criminology'. In Sollund, R. (ed.) *Global Harms: Ecological Crime and Speciesism*. New York: Nova Science Publishers, pp. 187–201.

South, N. (2014) 'Green criminology: Reflections, connections and harms', *International Journal for Crime, Justice and Social Democracy*, 3(2): 6–21.

Stretesky, P.B., Long, M.A. and Lynch, M.J. (2013) *The Treadmill of Crime: Political Economy and Green Criminology*. London: Routledge.

Sykes, G.M. and Matza, D. (1957) 'Techniques of neutralization: A theory of delinquency', *American Sociological Review*, 22(6): 664–670.

Tilman, D., Fargione J., Wolff, B., D'Antonio, C., Dobson, A., Howarth, R., Schindler, D., Schlesinger, W.H., Simberloff, D. and Swackhamer, D. (2001) 'Forecasting agriculturally driven global environmental change', *Science*, 292(5515): 281–284. Accessed on 25 November 2014 at http://www.sciencemag.org/content/292/5515/281.full.

United Nations Environment Programme (UNEP) (2012) 'Growing greenhouse gas emissions due to meat production', http://na.unep.net/geas/getUNEPPageWithArticleIDScript.php?article_id=92, http://www.unep.org/pdf/unep-geas_oct_2012.pdf.

Walters, R. (2010) 'Toxic atmospheres air pollution, trade and the politics of regulation', *Critical Criminology*, 18(4): 307–323.

Walters, R. (2013) 'Air crimes and atmospheric justice'. In South, N. and Brisman, A. (eds) *Routledge International Handbook of Green Criminology*. London: Routledge, pp. 134–150.

Walters, R. and Martin, P. (2014) 'Crime and carbon trading'. In Sorvatzioti, D., Antonopolous, G., Papanicolaou, G. and Sollund, R. (eds) *Critical Views on Crime, Policy and Social Control*. Cyprus: University of Nicosia Press, pp. 170–183.

Walters, R., Westerhuis, D. and Wyatt, T. (eds) (2013) *Emerging Issues in Green Criminology: Exploring Power, Justice and Harm*. Basingstoke: Palgrave Macmillan.

Warchol, G.L. (2004) 'The transnational illegal wildlife trade', *Criminal Justice Studies*, 17(1): 57–73.

Wellsmith, M. (2010) 'The applicability of crime prevention to problems of environmental harm: A consideration of illicit trade in endangered species'. In White, R. (ed.) *Global Environmental Harm: Criminological Perspectives*. Devon: Willan, pp. 132–149.

White, R. (2007) 'Green criminology and the pursuit of social and ecological justice'. In Beirne, P. and South, N. (eds). *Issues in Green Criminology*. Devon: Willan.

White, R. (2008) 'Crimes against nature'. *Environmental Criminology and Ecological Justice*. Devon: Willan Publishing.

White, R. (2011) *Transnational Environmental Crime: Toward an Eco-global Criminology*. London: Routledge.

White, R. (2012) 'The foundations of eco-global criminology'. In Ellefsen, R., Sollund, R. and Larsen, G. (eds) *Eco-global Crimes. Conteporary Problems and Future Challenges*. Farnham: Ashgate.

White, R. (2013a) *Crimes Against Nature: Environmental Criminology and Ecological Justice*. London: Routledge.

White, R. (2013b) 'The conceptual contours of green criminology'. In Walters, R., Westerhuis, D.S. and Wyatt, T. (eds) *Emerging Issues in Green Criminology: Exploring Power, Justice and Harm*. Basingstoke: Palgrave, pp. 17–33.

Whyte, W.F. (1943) *Street Corner Society*. Chicago: University of Chicago Press.

Woodroffe, R., Thirgood, S. and Rabinowitz, A. (eds) (2005) *People and Wildlife, Conflict or Co-Existence?* (No. 9). Cambridge: Cambridge University Press.

Wyatt, T. (2013) *Wildlife Trafficking: A Deconstruction of the Crime, the Victims, and the Offenders*. Basingstoke: Palgrave Macmillan.

Wyatt, T., Beirne, P. and South, N. (2014) 'Guest editors' introduction', *International Journal for Crime, Justice and Social Democracy, Special Edition: Green Criminology Matters*, 3(2): 1–5.

Young, J. (2007) *The Vertigo of Late Modernity*. London: Sage.

Ystehede, P.J. (2012) 'Constructing a meta-history of eco-global criminology: On brute criminologists, mortified bunnies, nature and its discontents'. In Ellefsen, R., Larsen, G. and Sollund, R. (eds) *Eco-global Crimes: Contemporary Problems and Future Challenges*. Farnham: Ashgate Publishing, pp. 57–71.

Zaffaroni, E.R. (2013) *La Pachamama y el humano*. Ediciones Madres de Plaza de Mayo: Ediciones Colihue.

Zalik, A. (2010) 'Volatilidad y mediación en diferentes campos petroleros: Las Arenas Bituminosas y el Delta del Níger como lugares en controversia', *Umbrales, Special Issue: Hidocarburos, Politica y Sociedad*, 20. CIDES: Bolivia, pp. 307–336. http://www.cides.edu.bo/webcides/images/pdf/Umbrales_20.pdf.

Zalik, A. (2012) 'Legal oil, ethical oil and profiteering in the Niger Delta and the Canadian North', *Arts, Activism, Education, Research*. Accessed on 30 September at http://platformlondon.org/2012/01/30/legal-oil-ethical-oil-and-profiteering-in-the-niger-delta-and-the-canadian-north/.

2

State–Corporate Environmental Harms and Paradoxical Interventions: Thoughts in Honour of Stanley Cohen

Avi Brisman and Nigel South

Introduction

Stanley Cohen was concerned with the crimes and harms that are perpetrated by legitimate bodies, such as states and corporations, and with the way in which these acts and omissions, their impacts and consequences (discussed further below), have been routinely ignored, overlooked, excused or simply denied. In various influential works, such as his 1993 article, 'Human rights and crimes of the state: The culture of denial', his 1985 book, *Visions of Social Control*, and his 2001 book, *States of Denial: Knowing about Atrocities and Suffering*, Cohen appealed to us to open our eyes, see and acknowledge the hidden crimes, horrors and indignities inflicted by humans on others, and his work explored important themes of truth and deception: the distortion of the former and the ways in which we produce the latter individually and collectively. The victims of ignored or almost invisible crimes and harms can easily be overlooked when offenders seek to hide their actions and the injuries caused, and when these victims are already socially invisible, marginalized or forgotten (Davis et al., 2014; Hall, 2014). No great effort at camouflage or disguise is required if the perpetrators or conspirators can enlist the willing cooperation of many or most in buying into the cover-ups, the denials and the comfortable avoidance of challenge. Stealthy misdirection, misinformation and the power to pay for legal harassment and media control shape a socioeconomic landscape in which the crimes and harms for which the powerful are responsible continue much as ever (Brisman, 2012). This chapter considers some examples of such crimes and harms from recent decades, focusing on environmental degradation and destruction, and taking as a template the spirit and insights of Cohen's work.

At the core of much of Cohen's writing – especially in *Visions of Social Control* and *States of Denial* – made more explicit in the latter but also present in the earlier book – was a concern for the protection of human rights. As Weber and colleagues (2014: 78) observe,

> No criminologist has pursued the project of aligning critical criminology with human rights activism with such effectiveness and vigour as the late Stanley Cohen (see Downes et al., 2007). His contributions include seminal works on state crime ... and the sociology of denial ..., a period spent 'doing human rights' with a non-governmental organisation ..., and active engagement with UN human rights mechanisms ...

It is of interest to us here, however, that Cohen did not extend his attention to the rights of other species or the state of the wider environment. He explains this in the following way: 'The concept of compassion fatigue may be a little shaky', he says of the idea that 'each new moral demand makes coping harder,' but he observes that he tested this proposition on himself 'by looking at' his 'own reactions to environmental and animal rights issues'. The result, he reported, was that although unable to

> find strong rational arguments against either set of claims ... emotionally, they leave me utterly unmoved. I am particularly oblivious—in total denial—about animal issues.... in the end, much like people throwing away an Amnesty leaflet, my filters go into automatic drive: this is not my responsibility; there are worse problems; there are plenty of other people looking after this.
>
> (Cohen, 2001: 289)

This chapter adopts a green criminological perspective and we start with an outline of this and how it would, and also would not, fit with Cohen's own priorities.

Cohen, anti-criminology and a green perspective

A 'green perspective' for criminology seeks neither to propose a definitive theory with respect to the causes of environmental crime or harm nor offer a specific set of solutions, but presents the argument that criminology should be more sensitive to the extent and implications of these urgent and globally important matters, from pollution and its regulation, through species extinction to the implications of climate change (Brisman, 2014; South, 1998, 2014; South and Brisman, 2013). Importantly, green criminological research is concerned with that which is illegal as proscribed by law and defined as 'crimes' as well as 'harms' that are not legally defined and addressed (Brisman, 2008). We suggest that these concerns and problems connect with those with which Cohen engaged because, as Gianolla (2013: 64) argues, 'humans cannot be protected without protecting Nature—and . . . this truth is particularly stark in the contemporary context. Indeed, it can be argued that the failure of mainstream human rights to embrace Nature's significance now has direct and negative implications for the capacity of mainstream human rights to fulfil their mandate.' We hope that Cohen would have had more sympathy with this position than the view of green criminology which traditional criminology may take, which Potter (2013: 125) suggests would run as follows: 'the perceived focus of green criminology—prioritising a concern for nature over social problems and often moving well beyond legal definitions of "crime"—leads to accusations that it is not really the proper business of *criminology* at all.' We would note, however—and we believe that Cohen would too—that this is not a new dilemma for criminology. As such, we agree with Ruggiero (2013: 261) when he argues that just as it has been necessary to recognise the 'continuity between legality and illegality' in order to examine 'white-collar, corporate, state crime, and . . . the crimes of the powerful in general', this same approach can

> be fruitfully utilised when environmental harm is analysed because such harm may be produced by both criminal conduct and completely lawful initiatives. The study of environmental harm, in other words, presents the same theoretical predicament experienced by students of the crimes of the powerful and campaigners mobilising against them. As Sutherland (1949) realized, research on the crimes of the powerful is difficult without a willingness to expand one's sample well beyond the legal definitions of crime.

The adoption of a perspective that embraces 'harms' as an alternative to the state-defined nature of 'crimes' is in keeping with the arguments and critical trajectory of the work of Cohen (1993, 2001), as well as

Schwendinger and Schwendinger (1970), Hillyard and Tombs (2004) and others (e.g. Potter, 2013: 134).

That said, Cohen was, of course, a major critic of the field of 'criminology', yet he engaged with its ideas and practice frequently and profoundly in search of an 'intellectual perspective that lies outside the ideology and interests of those who run the crime-control system and the academics they hire to help them' (Cohen, 1988: 9). As part of a long-running, arms-length appraisal of the strengths and weaknesses of 'criminology', one assumption that he interrogated was the idea that 'criminology' and its theories and proposals travel well—in other words, that its concerns, methods, theories, analyses and conclusions are transferable across contexts. The alternative view is that there will always be something (or a great deal) that is inevitably 'lost in translation'. As Bowling (2011: 362–363) points out, the basis of this debate is that Western criminology can be 'criticized because its theoretical presumptions are often misleading when applied to other contexts, miss the point, or are unhelpful in other ways (Cain, 2000)'.

Recognizing the relative neglect of comparative efforts in the past, criminology has recently become more avowedly globalized in its concerns and intellectual sphere of activity, although this expansion of horizons is not without its own issues or complexity. The breadth and scope of an expanded remit presents its own challenges. For example, Bowling (2011: 363) outlines three approaches—the comparative, the transnational and the global—in the following ways: all share some features and aim to address a common 'ethnocentric myopia', with the comparative perspective seeking to 'compare one place with another', the transnational perspective studying 'linkages between places' and a global criminology aspiring to 'bring together transnational and comparative research from all regions of the world to build a globally inclusive and cosmopolitan discipline'. In the development of a 'green criminology', a global or planetary perspective is central. For example, White's (2011: 2013) proposal of an 'eco-global criminology' is an explicit statement of this relevance, while South (1998: 226), in one of the first pieces on 'green criminology', noted that the exploration of green issues in criminology would also be relevant to the study of other emerging topics of international importance, such as crimes of war and violations of human rights. The development of empirical work in green criminology has also readily encompassed a global perspective as being essential for the study of transnational transfers, trafficking and circulation of waste, wildlife, pollution and so on (Bisschop, 2012a, 2012b; Sollund, 2013a, 2013b; Wyatt, 2013a, 2013b; Walters, 2010).

In following Cohen's path, we consider crimes and harms that are caused by states and corporations, as well as processes and implications of denial. We diverge, however, by shifting our focus from humanitarian to environmental concerns—although the two are absolutely intertwined (see, e.g., Brisman, 2004, 2007). We start by examining examples of state–corporate crimes that affect and abuse the environment, including human and non-human populations. These examples are also considered in terms of the denial of their importance, impact or even occurrence. In this respect, where the exercise of power by stealth leads to invisibility or indifference, where abuse and exploitation are hidden from view (e.g., Sollund, 2008; Vialle, 1994), it is hard to apply principles of remedy, justice or rights. Green criminology has critically examined such discourses and representation concerning environmental matters, including techniques of neutralization, of minimization of significance, and of outright denial, which we further illustrate in the sections below.

State–corporate crimes and environmental harms

Kramer and colleagues (2002: 271; see also Kramer, 2013: 157; Kramer and Michalowski, 2012: 76; Michalowski and Kramer, 2006: 15) define 'state–corporate crimes' as 'illegal or socially injurious actions that result from a mutually reinforcing interaction between (1) policies and/or practices in pursuit of the goals of one or more institutions of political governance and (2) policies and/or practices in pursuit of the goals of one or more institutions of economic production and distribution'. In line with the second element of this relationship, corporate interests will aim to increase their competitiveness and secure their access to particular markets and resources, and this may involve lowering environmental standards and undermining or violating the rights of activists who are seeking to protect the environment (see, e.g., Stretesky et al., 2013, 2014). In this, they will be supported by the first element of the relationship whereby institutions of political governance will permit, or deny the significance of, these actions. This is the *modus operandi* of a neoliberal discourse of 'growth and freedom' that has become dominant in recent decades and that has intensified the potential destructiveness of corporate enterprise—usually with the approval ('it is for the greater good') or denial ('there is no evidence of real damage') of the state. This mutually reinforcing neoliberal relationship of corporation and state is also clear when, in the midst of a global economic downturn, remedies to mitigate prevailing economic insecurity and depression are viewed almost entirely in terms of increasing production and consumption (Ruggiero and South, 2013). As White (2013: 245–246) argues,

> Economic efficiency is measured in how quickly and cheaply commodities can be produced, channeled to markets, and consumed. It is a process that is inherently exploitive of both humans and nature. The net result is a series of interrelated trends…: *Resource depletion*—extraction of non-renewable minerals and energy without development of proper alternatives, as well as over-harvesting of renewable resources…; *Disposal problems*—waste generated in production, distribution and consumption processes; pollution associated with transformations of nature…; *Corporate colonisation of nature*—genetic changes in food crops, use of plantation forestry that diminishes bio-diversity, and preference for large-scale, technology-dependent and high yield agricultural and aquaculture methods that degrade land and oceans and affect species development and wellbeing; [and] *Species decline*—destruction of habitats, the privileging of certain species of grain and vegetable growing over others for market purposes, and the super-exploitation of specific plants and animals due to corporate-defined consumer taste and mass markets.

Consumption is a key feature here, and Agnew (2013: 58) sees the ordinary acts of consumption within everyday life as cumulative contributors to 'ecocide—or the contamination and destruction of the natural environment in ways that reduce its ability to support life'. As Agnew (2013: 58) explains, these ordinary acts have several characteristics: 'they are widely and regularly performed by individuals as part of their routine activities; they are generally viewed as acceptable, even desirable; and they collectively have a substantial impact on environmental problems.' Of course, the idea that the routine activities of everyday life contribute to harm is likely to be met by familiar denials of responsibility, injury and of victims, condemnation of the environmentalist 'do-gooders', and appeals to the loyalty owed to self, family, friends and the market (Sykes and Matza, 1957), all representing pressing needs and expectations. Consumers will often prefer not to connect with wider environmental consequences (Brisman and South, 2014: 51–68). To come full circle, then, this is a problem not only of everyday practice (Bourdieu, 1990; Shove et al., 2012) but of the classic structure/agency variety, so as Fay (1996: 51) observes, 'explanations of the most individualistic appearing acts are a function of impersonal laws and forces characterising social wholes'. Nowhere does this 'social whole' of the state exert its impersonal and depersonalizing laws and forces more destructively on individual humans, other species and the natural environment than in the pursuit of war, to which we turn next.

The toxic impact of wars on the environment

As Carrabine and colleagues (2014: 463–464) argue, wars, invasions and colonialism are usually justified by states as being for the 'greater good' of civilization, although, in fact, they reflect more basic interests of state and corporate power, sometimes enacted with 'legitimacy', sometimes with a less secure legal basis. In the latter case, as Cohen (1993) shows, states engage in 'spirals of denial' in order to hide and forget their illegal activities. States may first try to employ 'total denial' by suggesting that an atrocity or similar offence did not actually occur, or may then engage in 'partial denial' by conceding that something did happen but 'what happened was really something else'. Examples of this include the USA's renaming of civilian deaths in the Iraq War as 'collateral damage', or the reframing of the use of torture in the War on Terror.

Military interventions justified in the name of peace, carried out by Western democracies and directed against those who abuse human rights and threaten genocidal actions, can also, with some irony, undermine and ignore both human and environmental rights, with the consequences being minimized, ignored or denied. In both of the Gulf Wars and in military interventions in Afghanistan and the Balkans, the use of depleted uranium shells led to damage that the Australian Medical Association for Prevention of War argued

> cannot be contained to 'legal' fields of battle; they continue to act after the conclusion of hostilities; they are inhumane because they place the health of non-combatants, including children and future generations, at risk; and they cannot be used without unduly damaging the natural environment.
>
> (2003, quoted in White, 2008: 34)

Perhaps the most devastating example of environmental destruction that has been pursued as a tactic or tool of war against human enemies remains the use of Agent Orange (AO) by US forces in the Vietnam War (see Brisman et al., 2015). The 'toxic remnants of war' (TRW), however, can be found in many other cases, and they give rise to serious humanitarian and environmental concerns. Ghalaieny (2013: 2) defines TRW as 'Any toxic or radiological substance resulting from military activities that forms a hazard to humans and ecosystems' and cites the dioxin contamination from the use of AO as a 'key example of the need for a formalised mechanism to deal with conflict-related pollution'. And, of course, denial is part of this story. As Ghalaieny (2013: 2) explains,

> US authorities were aware that the AO was contaminated with dioxin at the time but continued to use it on the basis of the military advantage they felt defoliation offered. Exposure to dioxin has subsequently been implicated as a cause of the birth defects documented in Vietnamese civilians, yet the problem is only now being acknowledged and is still far from resolved, more than 40 years after the initial contamination.

Human rights and environmental concern are brought together in the TRW Project, which aims to promote a humanitarian-centred approach to military-origin contamination using 'Epidemiological studies, environmental assessment and evidence of exposure' to 'assist in resolving controversies regarding the environmental origins of disease. Unsurprisingly, any association with vested interests such as industrial or military activities can trigger controversy' (Ghalaieny, 2013: 2). As one example, the crimes and harms that lie at the door of the nuclear state are numerous, although many more will remain unknown due to the secrecy surrounding the industry (Kauzlarich and Kramer, 1998). The use of nuclear weapons, the use of extraordinary powers by police and security services to guard nuclear secrets, and proposals to build new nuclear power plants have all, post-Chernobyl, Fukishima and in the continued absence of suitable waste-disposal strategies, rightly, continued to produce much debate.

Walters (2007: 188) has provided one analysis of 'The range of risks associated with commercial enterprises in research, power production, telecommunications, medicine and pharmaceuticals as well as state activities in military defence and war [which] all utilise varying degrees of radioactive substances that produce waste.' And there lies (literally) the problem—what to do with lethal waste? The dumping of toxic radioactive waste at sea has been widely documented (Ringius, 2001; Parmentier; 1999) and yet of all the materials that humanity might choose to dispose of in the oceans, radioactive waste and functional but decommissioned reactors must surely be high on the list of dangerous substances. While radioactive waste may be recycled in some forms and can be exported legally or illegally, and while valuable for some purposes, it is a difficult commodity to manage. Hence, one response has been to simply dump it at sea—spectacular examples include the practice of the Russian navy which disposed of submarine reactors and nuclear waste in the Barents and Kara seas, and the exposure by Greenpeace in 2000 of a UK policy of the 1950s and 1960s to dump containers of nuclear waste near the Channel Islands. Approximately 28,500 corroding containers were discovered, and this was just one of many dump sites that were used before the banning of the practice (Walters, 2007: 189).

As Bauman (1989) and others have shown, war and its machinery provide one of the starkest ways in which modernity has demonstrated its ability to produce 'solutions' to 'problems'—whether with good or evil intent. In the same spirit of a search for 'solutions', modernity's engagement with 'policy problems' can produce paradoxes, ironies and unintended consequences. We highlight some of these in the next section.

The benign or malign benefits of modernity?

In 'Transnational criminology and the globalization of harm production', Bowling (2011) summarizes some of the key insights that Cohen expressed in his article, 'Western crime control models in the Third World: benign or malignant?' (Cohen, 1988). According to Bowling (2011: 365), Cohen argued that criminal justice practices often produced unintended negative consequences, and Cohen was especially 'critical of the application of damaging Western crime control models to developing countries'. As Bowling (2011: 365) explains, Cohen 'criticize[d] the tendency for criminologists from rich countries to export crime control theories to their poorer neighbours without the benefit of a comparative or a transnational perspective'.

Cohen's notion of the 'paradoxical counter productivity' of institutions and interventions—his belief that institutions, perhaps with the best of motives, may, with their interventions, produce 'disastrous results' (1988: 189)—offers some interesting parallels with the modernist export of models of production and consumption and some of their consequences. For example, Stretesky and colleagues (2013, 2014; Long et al., 2012) have applied 'treadmill of production' theory to an examination of how environmental harms are the direct result of Western-led processes of production, growth and capital accumulation, supported by the alignment of corporate and state interests. Similarly, Western-led consumption demands may cause socioeconomic and crime/conflict problems in developing nations that are the principal suppliers of raw materials and/or finished goods for this consumer market. There is a complex relationship here that has been noted in critical green criminology (Brisman and South, 2013; South and Brisman, 2012) but is also very interestingly reflected in a debate within environmental social sciences concerning what is termed the 'ecological modernization' thesis. This position is premised on the same kind of assumptions that underpin the modernizing impulse, which is accompanied by an ideological insistence on ways of 'doing things', and conditions of monetary loans and other aid that impose on developing nations a late-modern version of the missionary road to salvation, no longer via religion but now via Western economics, technologies and practices. As noted above, Cohen (1988) discussed such imposition in the case of Western crime-control models, raising questions of whether the outcomes were benign or malign. In the case of ecological modernization theory, critics have similarly questioned its central argument that ' "green" technologies—developed and commercialized in core nations—will benefit, or at least have the capacity to benefit, all people universally' (Bonds and Downey, 2012: 168), when, in fact, the dependence of green technologies on particular natural resources (e.g. rare metals), largely found in and extracted from developing or peripheral countries, means that they may benefit advanced consumer societies but generate 'environmental degradation, violence and social disruption in peripheral zones' (Bonds and Downey, 2012: 168; see also Kane and Brisman, 2013).

A similar set of issues and questions arise in relation to the 'Green Revolution' of the 1965–1980 period, 'in which high yielding crop varieties and an associated technology/policy package were exported via an international network of agricultural research institutes and donors to Asia and Latin America' (Brooks, 2005: 360), as well as with regard to the diffusion of new biotechnologies to developing nations in more recent decades. Both frameworks of intervention have been criticized, in ways that Cohen would have recognized, for reflecting 'neo-Malthusian reasoning about the relationship between population and food production, each proposing a new "technological fix" (Goodman and Redclift, 1991: 142) to intensify production (Nature, 2002); squeeze out small farmers and increase inequality (Shiva, 1991; Action Aid, 2003); and threaten biodiversity through the promotion of monocultures' (Brooks, 2005: 361 (internal footnote omitted)). For producer countries, the problems arising from these 'green' interventions are economic, ethical, humanitarian and environmental (Reed, 2002; Clark, 2013; Brisman and South, 2013). Corporations and states, however, do not believe that they necessarily have to actually change in fundamental ways to mitigate or reduce harms to the environment or contributions to climate change, so long as they can manage the risk (or the perception thereof). In other words, their strategy is one in which the management of risks of environmental damage or acceleration of climate change is preferable to tempering consumer demand or growth in production (Lefsrud and Meyer, 2012).

The creation of the 'need' for consumption produces profiteering based on the exploitation of the mineral, plant and wildlife wealth of developing nations. Global capital and the corporations of the developed world claim 'rights' to indigenous knowledge, patent the genetic or pharmaceutical products of plants of local origin, and pursue the mining of ores and felling of timber to meet the demand of global consumer markets in both advanced and emerging economies. But it is not only consumer products that carry a cost: despite denial in some quarters, the evidence shows that the planet is also incurring the cost of the abuse of the environment (see, e.g., Brisman, 2012; South, 2007; Wyatt, 2014).

Conclusion

Cohen filled the role of the critical—and globally sensitive—intellectual, exemplifying a commitment to ensuring that debate about matters such as values, morality and relativism continue rather than face dissipation or suppression. He feared that post-modernism's rejection of absolute values carries with it the danger of dismissing the universal relevance and applicability of norms or moral standards. According to the post-modern critique, Cohen (2001: 285) explained, values such as those reflected in the Universal Declaration of Human Rights could be seen as ethnocentric and individualistic—the global impositions of post-colonial but still dominant Western states. Yet, as he astutely and convincingly argued, to allow governments and others in positions of power to adopt a 'principled' rejection of the applicability of international human rights norms is to provide those with little or no regard for such rights with an 'official discourse of denial' with 'pernicious outcomes'. It is possible, as Cohen admitted (2001: 286), to caricature the highly complicated debate about cultural specificity versus universality (South and Weiss, 1998: 9–11). In the case of both human rights and environmental protection, however, there is a need to challenge and undermine the 'cultural alibis' and scientific denials that are employed to suggest that—whether the matter is corporate pollution, environmental victimization or climate change—'there is no problem here' and 'if there is a problem somewhere else then it is somebody else's problem'.

If human rights can lay claim to some sense of universality in the face of crimes such as genocide, then it is possible to conceive of a similar basis for principles and international law that could apply to crimes and harms against the environment. Higgins (2010; Higgins et al., 2013) has campaigned for the introduction of 'ecocide' as a fifth international crime against peace under a proposed amendment to the Rome Statute. This treaty, in force from 1 July 2002, established the International Criminal Court and set out its functions, jurisdiction and structure, but it also limited its remit to the investigation and prosecution of only the core international crimes—known as the four 'Crimes against Peace' (genocide, crimes against humanity, war crimes and the crime of aggression)—where states are unable or unwilling to pursue these themselves. The proposal to add a crime of 'ecocide' starts by defining it in the following way: 'Ecocide is the extensive damage to, destruction of or loss of ecosystem(s) of a given territory, whether by human agency or by other causes, to such an extent that peaceful enjoyment by the inhabitants of that territory has been severely diminished'.[1] We hope that Cohen would have supported this undertaking and the adoption of an international crime of ecocide to sit alongside the crime of genocide. There again, perhaps he would not, and it is our collective loss that we cannot pursue this debate with him.

Note

1. Were 'ecocide' to become the fifth international crime against peace, the International Criminal Court might restrict the interpretation of 'inhabitants' to 'humans'. On the other hand, the explicit reference to 'ecosystems' suggests that 'inhabitants' could be interpreted to include non-human species.

References

Action Aid (2003) *Going against the Grain*. London: ActionAid.
Agnew, R. (2013) 'The ordinary acts that contribute to ecocide: A criminological analysis'. In South, N. and Brisman, A. (eds) *Routledge International Handbook of Green Criminology*. London and New York: Routledge, pp. 58–72.
Bauman, Z. (1989) *Modernity and the Holocaust*. Cambridge: Polity Press.
Bisschop, L. (2012a) 'Is it all going to waste? Illegal transports of e-waste in a European hub', *Crime, Law and Social Change*, 58(3): 221–249.
Bisschop, L. (2012b) 'Out of the woods: The illegal trade in tropical timber and a European trade hub', *Global Crime*, 13(3): 191–212.
Bonds, E. and Downey, L. (2012) ' "Green" technology and ecologically unequal exchange: The environmental and social consequences of ecological modernization in the world-system', *Journal of World Systems Research*, 18(2): 167–86.
Bourdieu, P. (1990) *The Logic of Practice*. Cambridge: Polity Press.

Bowling, B. (2011) 'Transnational criminology and the globalization of harm production'. In Hoyle, C. and Bosworth, M. (eds) *What Is Criminology?* Oxford: Oxford University Press.

Brisman, A. (2004) 'Double whammy: Collateral consequences of conviction and imprisonment for sustainable communities and the environment', *William & Mary Environmental Law & Policy Review*, 28(2): 423–475.

Brisman, A. (2007) 'Toward a more elaborate typology of environmental values: Liberalizing criminal disenfranchisement laws and policies', *New England Journal on Criminal & Civil Confinement*, 33(2): 283–457.

Brisman, A. (2008) 'Crime-environment relationships and environmental justice', *Seattle Journal for Social Justice*, 6(2): 727–817.

Brisman, A. (2012) 'The cultural silence of climate change contrarianism'. In White, R. (ed.) *Climate Change from a Criminological Perspective*. New York: Springer, pp. 41–70.

Brisman, A. (2014) 'Of theory and meaning in green criminology', *International Journal of Crime, Justice and Social Democracy*, 3(2): 22–35.

Brisman, A. and South, N. (2013) 'Resource wealth, power, crime and conflict'. In Walters, R., Westerhuis, D. and Wyatt, T. (eds) *Debates in Green Criminology: Power, Justice and Environmental Harm*. Basingstoke: Palgrave Macmillan, pp. 57–71.

Brisman, A. and South, N. (2014) *Green Cultural Criminology: Constructions of Environmental Harm, Consumerism, and Resistance to Ecocide*, London and New York: Routledge.

Brisman, A., South, N. and White, R. (2015) 'Toward a criminology of environment-conflict relationships'. In Brisman, A., South, N. and White, R. (eds) *Environmental Crime and Social Conflict: Contemporary and Emerging Issues*. Surrey, UK: Ashgate, pp. 1–38.

Brooks, S. (2005) 'Biotechnology and the politics of truth: From the green revolution to the evergreen revolution', *Sociologia Ruralis*, 45(4): 360–379.

Cain, M. (2000) 'Orientalism, occidentalism and the sociology of crime', *British Journal of Criminology*, 40: 239–60.

Carrabine, E., Cox, P., Fussey, P., Hobbs, D., South, N., Thiel, D. and Turton, J. (2014) *Criminology: A Sociological Introduction*. London and New York: Routledge.

Clark, R. (2013) 'The control of conflict minerals in Africa and a preliminary assessment of the Dodd-Frank Wall Street Reform and Consumer Act'. In South, N. and Brisman, A. (eds) *Routledge International Handbook of Green Criminology*. London and New York: Routledge, pp. 214–229.

Cohen, S. (1985) *Visions of Social Control*. Cambridge: Polity.

Cohen, S. (1988) 'Western crime control models in the Third World'. In Cohen, S. (1988) *Against Criminology*, New Brunswick, NJ: Transaction Books (reprinted from Spitzer, S. and Simon, R. (eds) (1982) *Research in Law, Deviance and Social Control*, Vol. 4. Greenwich, CT: JAI Press, pp. 85–199).

Cohen, S. (1993) 'Human rights and crimes of the state: The culture of denial', *Australian and New Zealand Journal of Criminology*, 26(2): 97–115.

Cohen, S. (2001) *States of Denial: Knowing about Atrocities and Suffering*. Cambridge: Polity.

Davis, P. Frances, P. and Wyatt, T. (eds) (2014) *Invisible Crimes and Social Harms*. Basingstoke: Palgrave Macmillan.

Downes, D., Rock, P., Chinkin, C. and Gearty, C. (eds) (2007) *Crime, Social Control and Human Rights: From Moral Panics to States of Denial, Essays in Honour of Stanley Cohen*. Cullompton: Willan.

Fay, B. (1996) *Contemporary Philosophy of Social Science: A Multicultural Approach*. Oxford: Blackwell.

Ghalaieny, M. (2013) *Toxic Harm: Humanitarian and Environmental Concerns from Military-Origin Contamination*. Manchester, UK: Toxic Remnants of War Project. Accessed at http://www.toxicremnantsofwar.info/wp-content/uploads/2013/03/Toxic_Harm_TRWProject.pdf.

Gianolla, C. (2013) 'Human rights and nature: Intercultural perspectives and international aspirations', *Journal of Human Rights and the Environment*, 4(1): 58–78.

Goodman, D. and Redclift, M. (1991) *Refashioning Nature: Food, Ecology and Culture*. London and New York: Routledge.

Hall, M. (2014) *Interrogating Green Crime: An Introduction to the Legal, Social and Criminological Contexts of Environmental Harm*. Basingstoke: Palgrave Macmillan.

Higgins, P. (2010) *Eradicating Ecocide: Laws and Governance to Prevent the Destruction of Our Planet*. London: Shepheard-Walwyn.

Higgins, P., Short, D. and South, N. (2013) 'Protecting the planet: A proposal for a law of ecocide', *Crime, Law and Social Change*, 59(3): 251–266.

Hillyard, P. and Tombs, S. (2004) 'Towards a political economy of harm: States, corporations and the production of inequality'. In Hillyard, P., Pantazis, C., Tombs, S. and Gordon, D. (eds) *Beyond Criminology: Taking Harm Seriously*. London: Pluto, pp. 30–54.

Kane, S.C. and Brisman, A. (2013) 'Technological drift and green machines: A cultural analysis of the *Prius Paradox, CRIMSOC: The Journal of Social Criminology*, Green Criminology Issue (Autumn): 104–133. Accessed at http://socialcriminology.webs.com/CRIMSOC%202013%20Green%20Criminology.pdf.

Kauzlarich, D. and Kramer, R.C. (1998) *Crimes of the American Nuclear State: At Home and Abroad*. Boston: Northeastern University Press.

Kramer, R.C. (2013) 'Carbon in the atmosphere and power in America: Climate change as state-corporate crime', *Journal of Crime and Justice*, 36(2): 153–170.

Kramer, R.C. and Michalowski, R.J. (2012) 'Is global warming a state-corporate crime? In White, R. (ed.) *Climate Change from a Criminological Perspective*. New York: Springer, pp. 71–88.

Kramer, R.C., Michalowski, R.J. and Kauzlarich, D. (2002) 'The origins and development of the concept and theory of state-corporate crime', *Crime and Delinquency*, 48(2): 263–282.

Lefsrud, L. and Meyer, R. (2012) 'Science or science fiction? Professionals' discursive construction of climate change', *Organization Studies*, 33(11): 1477–1506.

Long, M., Stretesky, P., Lynch, M. and Fenwick, E. (2012) 'Crime in the coal industry: Implications for green criminology and treadmill of production', *Organization & Environment*, 25(3): 328–346.

Michalowski, R.J. and Kramer, R.C. (2006) *State-Corporate Crime: Wrongdoing at the Intersection of Business and Government*. New Brunswick, NJ: Rutgers University Press.

Nature (2002) 'Nature Insight: Food and the future', *Nature*, 418: 667–707.

Parmentier, R. (1999) 'Greenpeace and the dumping of waste at sea: A case of non-state actors' intervention in international affairs', *International Negotiation: A Journal of Theory and Practice*, 4(3): 435–457.

Potter, G. (2013) 'Justifying "green" criminology: Values and "taking sides" in an ecologically informed social science'. In Cowburn, M., Duggan, M., Robinson, A. and Senior, P. (eds) *The Value(s) of Criminology and Criminal Justice*. Bristol: Policy Press, pp. 125–141.

Reed, D. (2002) 'Resource extraction industries in the developing world', *Journal of Business Ethics*, 39(3): 199–226.

Ringius, L. (2001) *Radioactive Waste Disposal at Sea*. Cambridge: MIT Press.

Ruggiero, V. (2013) 'The environment and the crimes of the economy'. In South, N. and Brisman, A. (eds) *Routledge International Handbook of Green Criminology*. London and New York: Routledge, pp. 261–271.

Ruggiero, V. and South, N. (2013) 'Toxic state–corporate crimes, neo-liberalism and green criminology: The hazards and legacies of the oil, chemical and mineral industries', *International Journal for Crime, Justice and Social Democracy*, 2(2): 12–26.

Schwendinger, H. and Schwendinger, J. (1970) 'Defenders of order or guardians of human rights', *Issues in Criminology*, 5(2): 123–157.

Shiva, V. (1991) *The Violence of the Green Revolution: Third World Agriculture, Ecology and Politics*. London: Zed Books.

Shove, E., Pantzar, M. and Watson, M. (2012) *The Dynamics of Social Practice: Everyday Life and How It Changes*. London: Sage.

Sollund, R. (ed.) (2008) *Global Harms: Ecological Crime and Speciesism*. New York: Nova Science.

Sollund, R. (2013a) 'Animal trafficking and trade: Abuse and species injustice'. In Walters, R., Westerhuis, D.S. and Wyatt, T. (eds) *Emerging Issues in Green Criminology: Exploring Power, Justice and Harm*. Basingstoke: Palgrave Macmillan, pp. 72–92.

Sollund, R. (2013b) 'The victimization of women, children, and non-human species through trafficking and trade: Crimes understood through an ecofeminist perspective'. In South, N. and Brisman, A. (eds) *Routledge International Handbook of Green Criminology*. London and New York: Routledge, pp. 317–329.

South, N. (1998) 'A green field for criminology? A proposal for a perspective', *Theoretical Criminology*, 2(2): 211–234.

South, N. (2007) 'The "corporate colonisation of nature": Bio-prospecting, bio-piracy and the development of green criminology'. In Beirne, P. and South, N. (eds) *Issues in Green Criminology: Confronting Harms against Environments, Humanity and other Animals*. Cullompton: Willan, pp. 230–247.

South, N. (2014) 'Green criminology: Reflections, connections, horizons'. *International Journal of Crime, Justice and Social Democracy*, 3(2): 5–20.

South, N. and Brisman, A. (2012) 'Critical green criminology, environmental rights and crimes of exploitation'. In Winlow, S. and Atkinson, R. (eds) *New Directions in Crime and Deviance*. London and New York: Routledge, pp. 99–110.

South, N. and Brisman, A. (eds) (2013) *Routledge International Handbook of Green Criminology*. London and New York: Routledge.

South, N. and Weiss, R. (1998) 'Crime, punishment and the "state of prisons" in a changing world'. In Weiss, R. and South, N. (eds) *Comparing Prison*

Systems: Toward a Comparative and International Penology. Gordon and Breach: Amsterdam, pp. 1–18.

Stretesky, P.B., Long, M.A. and Lynch, M.J. (2013) 'Does environmental enforcement slow the treadmill of production? The relationship between large monetary penalties, ecological disorganization and toxic releases within offending corporations', *Journal of Crime and Justice*, 36(2): 233–247.

Stretesky P., Long, M. and Lynch, M. (2014) *The Treadmill of Crime: Political Economy and Green Criminology*. London: Routledge.

Sutherland, E. (1949) *White Collar Crime*. New York: The Dryden Press.

Sykes, G.M. and Matza, D. (1957) 'Techniques of neutralization: A theory of delinquency', *American Sociological Review*, 22(6): 664–670.

Vialle, Nöelie (1994) *Animal to Edible*. Cambridge: Cambridge University Press.

Walters, R. (2007) 'Crime, regulation and radioactive waste in the United Kingdom'. In Beirne, P. and South, N. (eds) *Issues in Green Criminology*. Cullompton, Devon, UK: Willan.

Walters, R. (2010) 'Eco crime'. In Muncie, J., Talbot, D. and Walters, R. (eds) *Crime: Local and Global*. Cullompton, Devon, UK: Willan.

Weber, L., Marmo, M. and Fishwick, E. (2014) *Crime, Justice and Human Rights*. London: Palgrave.

White, R. (2008) 'Depleted uranium, state crime and the politics of knowing', *Theoretical Criminology*, 12(1): 31–54.

White, R. (2011) *Transnational Environmental Crime: Toward an Eco-global Criminology*. London and New York: Routledge.

White, R. (2013) 'Eco-global criminology and the political economy of environmental harm'. In South, N. and Brisman, A. (eds) *Routledge International Handbook of Green Criminology*. London and New York: Routledge, pp. 243–260.

Wyatt, T. (2013a) 'Uncovering the significance of and motivation for wildlife trafficking'. In South, N. and Brisman, A. (eds) *Routledge International Handbook of Green Criminology*. London and New York: Routledge, pp. 303–316.

Wyatt, T. (2013b) *Wildlife Trafficking: A Deconstruction of the Crime, the Victims, and the Offenders*. Basingstoke: Palgrave Macmillan.

Wyatt, T. (2014) 'Invisible pillaging: The hidden harms of corporate biopiracy'. In Davis, P., Frances, P. and Wyatt, T. (eds) *Invisible Crimes and Social Harms*. Basingstoke: Palgrave Macmillan, pp. 161–177.

3

Looking into the Abyss: Bangladesh, Critical Criminology and Globalization

Wayne Morrison

Introduction: Bangladesh and the Abyss

'Bangladesh stares into abyss after bloody vote "farce" ': so The Voice of Russia on 6 January 2014 headlined its coverage of a disputed and 'blood-soaked' national election where the ruling Awami League 'romped to victory'. The election followed a tumultuous 2013 and had been boycotted by the opposition alliance parties, which had called their supporters onto the streets in the tradition of 'hartals' (politically motivated strikes that are designed to make the situation ungovernable), which resulted in considerable violence – several hundred individuals who were linked to various parties had been killed or 'disappeared' in the build-up, and over 200 polling stations were firebombed on polling day. Few gave the elections any validity as a reflection of the general will of the people, but some wondered if the result – constitutionally a valid re-election – reflected an act of brazen politics by a government whose mandate in the 2008 election had focused upon a promise to establish domestic war crimes trials to push for justice and end impunity, but who, many suspected, had turned that process into a political vendetta. 2013 was a year of mass demonstrations in Bangladesh, for working wages in the garment industries, against the apparent legal immunity of factor owners who operated factories in breach of health and safety rules but most particularly in favour or against its war crimes trials. These were termed 'domestic' trials of individuals for 'international crimes' that were committed in the bloody birth of Bangladesh in 1971 (the War of Liberation), and the act that set up the tribunal had specifically stated that the criteria for judging them should be the standards

43

of what Bangladesh could afford. They were never intended to be the highly expensive (what some in Bangladesh termed 'luxury') trials of the International Criminal Court, the ad hoc tribunal in the Hague for the ex-Yugoslavia, or the hybrid tribunal in Cambodia. That they could in part fall short of international standards was not in itself cause for concern for much of the Bangladeshi public. However, as the major defendants were political opponents of the governing Awami League, a greater concern was that it criminalized politics even more and effectively turned what should have been a legal process into a political one.[1] 2013 was meant to finally give justice to the victims, to reassure all that the past could be got to grips with, and to demonstrate that a committed government could achieve justice. Conversely, many predicted that if the opposition won the elections of early 2014, the trials would end and any verdicts would be annulled. So the question arises: Was the outcome, the demonization of those who engaged in violent protests against it and a subsequent election boycotted by the opposition, unintended or brilliantly devious?[2] Throughout 2014 the government looked dominant, with many of its Jamaat-e-Islami (the main Islamic party) opponents in jail or having been killed, and its 'secular' opposition sidelined by not participating in the elections (thereby losing their parliamentary positions and associated power, while some languished in jail awaiting trial on corruption charges) and unable to mount the traditional street demonstrations in the face of police and paramilitary operations. Bangladesh was experiencing a constitutional one-party rule that seemed to use the clothing of a rule of law to punish its opposition and allow it to use open and hidden force by making constant reference to the threat of terrorism, and to those who criticize it as betraying the legacy of the War of Liberation and the sacrifices of the 'martyrs' who had died in it. However, a new and brutal cycle of political violence began when on 3 January 2015 police prevented Khaleda Zia, the leader of the opposition Bangladesh Nationalist Party, from leaving her office and the police then subsequently refused to give permission to allow the party to organize a rally it had planned on 5 January 2015, the anniversary of the elections. Khaleda Zia then announced a non-stop 'blockade' of the country, and this resulted in the arrest of a number of senior BNP leaders. As of mid-March over 120 had died, many as a result of petrol bombs being thrown at buses, over 8,000 had been arrested and there were numerous claims of individuals 'disappearing' (see the political blog of David Bergman at http://bangladeshpolitico.blogspot.com).

I presume that the editors of The Voice of Russia did not intend to reference Nietzsche's famous epigram: 'Whoever fights monsters should see to it that in the process he does not become a monster. And remember when you look into the abyss that the abyss also looks into you' (146 in Nietzsche's *Beyond Good and Evil*), but I will write as if they did.

Criminology and travel: Remembering theoria

This chapter is inherently personal; it considers what it means as a 'critical criminologist' to travel to and contemplate Bangladesh and the events of 2013/2014. My organizing idea is that of an abyss and Nietzsche's radical (and usually termed anti-reason) claim that 'is seeing itself – not seeing abysses?' (1961: 177). Certainly this may seem a strange concept for a criminologist. Today criminological theory is largely performative; activity is focused on scoping out and building up a research project, then following the terms of the grant that one has obtained, with great attention being given to building on the methodologies of the canon. In my own history of criminological theorizing (Morrison, 2006), I emphasized how so much of criminology has been a labouring for security in what I termed 'civilised space', what we might visualize as the protected internal territory of Hobbesian sovereignty. Outside, the external, the colonial, waging of war did not provide material for the criminological gaze. But the modern term 'theory' comes from 'theoria', which meant to look at, to see, to contemplate, and classic narratives of theoria said that one needs to be taken away from one's habitual surroundings to see afresh (or in Plato's story of the cave, taken out to see truth in the light of the sun, after which one must return in knowledge of the truth to rule). In related form, to contemplate led onto a process of contemplation, which meant to attain knowledge of the divine, and that direct experimental knowledge thereafter structured one's life. The religious meaning of contemplation gave an experience of finding foundations from which to interpret what one saw with the help of a fundamental presence that guided one. That now seems for most of us an understanding from another time and place – we tend to assume a secular subject and a naturalist object of study: man as a rational (or emotive) calculator, a subject of subjective desiring. But it is as well to remember that criminology has always been a particularly visual science that is linked to the theme of replacing conflicting and irrational interpretations of the sacred by enlightenment, and images of the Goddess of Justice having her eyes opened to reality (borrowing classic images of the enlightenment as a new dawn, where on the horizon rose the herald sun, the rosy aurora; truth rejoices and false wisdom is horrified). Consider the frontispiece of a German 1788 edition of Beccaria's *On Crime and Punishment* (first published anonymously in 1764 in Milan) where a clear-sighted (i.e. not blindfolded) God-dess of Justice turns away from the offering of severed heads that an executioner holds out. Beccaria supplied the outline drawing, and the contrast with Bangladesh and the massive protests in 2013 demand-ing the death penalty for those convicted in the war crimes trials is instructive. Beccaria's condemning of torture and the death penalty puts into play what David Garland called the 'governmental project' for criminology, and Beccaria's premises of reason, utility and deter-rence resulted in his rejection of executions. Justitia holds up a hand to the executioner's offering, indicating no, and turns her eyes away in disgust; at her side the scales of justice are entangled with tools that are used in farming and industry; in the background we see rolling hills lit by a tranquil sky. Consider also Lombroso's famous moment of realization: 'At the sight of that skull, I seemed to see all of a sud-den, lighted up as a vast plain under a flaming sky, the problem of the nature of the criminal –' (Lombroso, full extract Morrison, 1995: 124). Garland terms this the other, individualist side of criminology, the search for the criminal man, the search for criminality. The history of criminology could be written in terms of searching for its Aurora (Latin) or Eos (Greek), the goddess of the morning red, who brings up the light of dawn from the east and allows us to see in that light. In committed criminological writings one has a feel for the aurora that accompanies the writer/speaker; administrative criminology replaces the aurora with methodology (Young, 2011).

What guides critical criminology? A critical criminologist looks at the world differently from a mainstream or administrative criminologist. A caveat: I am not an anti-positivist. Along with positivism I hold to the existence of an external reality and that the basis of our 'science' is the observation of that 'reality'. Criminological positivism, however, is essentially lazy – it holds too narrow a world of facts, does not subject the items that its vision brings forth to contextual interpreta-tion, does not seek to deconstruct its grounding but accepts its role as applied science working within a theoretical/political synthesis that it often notes implies that 'crime' is a social construction but then acts as if that understanding was irrelevant. It is also Eurocentric. While we might agree that the foundational distinction is between 'crime' and 'politics' (Mannheim, 1965), when we stand outside the Western main-stream it is harder to spot the difference (Cohen, 1996). If we then apply this and deconstruct criminological commonsense, where do we stand? My answer here is that we do indeed get an abyss – an abyss that must be understood theoretically and practically.

An abyss conjures up notions of a vast chasm, a lack of foundations, bottomlessness; the opposite of the regulated normality, the civilized space of (post-Hobbesian) sovereignty. References to an abyss were common in a year for Bangladesh which had seen a number of garment factory fires or building collapses (with over 1,129 bodies recovered from one such building), and huge demonstrations for and against its war crimes trials process, which in fact appeared to target Islamic opposition politicians. When, in February 2013, the first sentence given to a person in custody was announced, that of life imprisonment, huge protests demanding the death penalty (called the Shabagh movement and involving social media activists) took place (see Ashraf, 2013, on the paradox of 'bloodthirsty non-violence'). These in turn brought out even bigger protests that were organized by the Islamic parties, and in the government crackdown, claims of massacres and counterclaims of Islamic fundamentalist terrorism were rife. For a while it looked as if the government would lose popularity, be exposed as shamelessly parading justice while doing vengeance, and would lose any ability to govern. However, along with a refusal to allow a caretaker administration, the government 'succeeded'.

Encountering Bangladesh, reading criminology: A personal reflection

I am an academic with a long record of university administration, always living a duality of theoretical aspiration and practical routines and duties. I first went to Bangladesh in 1997 on University of London business. A small private teaching institution, run by Bangladesh's first female barrister, wanted to link with the university's external programme for law (where students sit the University of London law examinations but are not registered on campus at one of the London colleges), so a two-man 'inspection team' was sent. It was assumed that this would be a wasted trip. The airport was a shock: one crossed borders geographically and existentially, thousands lined the wire walls outside clamouring for a view, many were disabled calling for alms, we were escorted by armed police into the hotel van and then, as we went to the hotel, tens of thousands lined the traffic-jammed streets throwing coloured water everywhere.[3] Bangladesh had defeated Scotland at cricket and had made it to the World Cup 50 overs cricket competition: sights of joy and despair. Later that evening, after the first stage of our official meetings, in a fit of overconfidence, I left the protected hotel and took a rickshaw to what I thought I had seen as a set of coffee shops. A few hundred yards further the rickshaw turned into a side street and some 20 or so other rickshaws and men surrounded us: I became temporarily kidnapped. As a Western male it was assumed that I had no interest in coffee but only women: for the fee of US$ 100 I would be 'escorted' to a brothel and expected to 'sample'. At first this was a curious stand-off: I was physically fit and taller than any of my opponents but they far outnumbered me and might carry knives. A contract was agreed: I would at least look at the brothel and then hand over all of the money that I had on me. So I visited a building of appalling extremes: women crammed into small wooden boxes, some eyes visible with deadened gaze. Almost physically sick, I turned and threw some US$ 15 on the ground, then after some 'negotiations' I turned my pockets inside out to show that any fight would be for nothing and they agreed to take me back to the hotel for the promise of another US$ 85 that I would bring down from my room. Once inside the hotel I reneged on the 'contract': so ended my first day.[4]

On our return to London, to the surprise of many we recommended that the external programme should recognize the institution and we gave our opinion that there could be great promise in Bangladesh. The people whom we had met talked about globalization and opening up to the world. We were moved by their energy in the midst of seemingly chaotic officialdom.

Subsequently, in 1999, I became director of the University's external programme for law and thenceforth visited Bangladesh at least once a year. I learnt of its War of Liberation and used it as a forgotten example of genocide in my writings (see, e.g., 2006: Chapter 9). I made strong friends with a devout Muslim family, which was headed by one of the student leaders of that war, and I would usually stay with them with the privileges (and obligations) of a family that employ over 15,000 workers in garment and other businesses. I have visited several garment and other factories (on two occasions 'posing' as an inspector from the International Labour Organization), and in the course of a now successful programme with around 2,000 students studying law, I have met with numerous government ministers, chief justices and attorneys, and have attended many weddings. In 2013 and 2014 I engaged in 'unstructured' research on the war crimes trials (see Morrison, 2013), meeting some of the judges, members of the prosecution, the defence team and witnesses, as well as observers, and wading through judgements of hundreds of pages. The result has been to encounter a confusion of discourses, narratives and claims. In my practical work connected to Bangladesh, expanding the programme and lecturing to groups of students, I felt part of a process of modernization, of giving languages and tools for an emerging middle class (many of whom were young women), and in my lectures I tried to offer some of the language and concepts as resources to interpret these students self-positioning in Bangladeshi society. Yet when I sought definite answers in the intellectual resources that I carried, I was tentative, searching and always in danger of becoming lost. For critical criminologists, 2013–2014 was a time when we marked with respect and regret the passing of Stan Cohen, Jock Young and then Stuart Hall. In Bangladesh it was as if some of their work had been read and applied in converse terms to the author's intentions, and for other parts that Bangladesh was a place 'other', beyond their consideration, almost a counter-factual.

Consider Jock Young

In 2001 I read Jeremy Seabrook's *Freedom Unfinished: Fundamentalism and Popular Resistance in Bangladesh Today* (Seabrook, 2001), in which he articulated an interpretation of the dichotomy of images that assailed both him and me. On one side, 'extremes of poverty, political corruption and violence, dependence on international aid, child labour ... the issue of population', which he called 'that sinister distortion of the country's greatest resource and only real hope – its people'; on the other, the courage and endurance of the poor, their inventiveness and creativity, and I also marvelled with him at the 'miracles of ingenuity and effort' by which currently (2014) about 168 million people survive on 67,000 square miles (i.e. the size of England and Wales). But Seabrook's analysis was ultimately chilling:

> The ravages of nature, political violence, corruption – all conceal the fact that Bangladesh is still in the grip of a continuing and indeed undeclared civil war, the unfinished business of its creation in 1971. Bangladesh was lost to Pakistan then; a loss which Pakistan and a powerful minority in Bangladesh never accepted. What was lost to Pakistan may still be regained for Islam, but not the traditional fluid and tolerant Islam of Bengal.
>
> (Seabrook, 2001: 3)

While I found his experiences increasingly resonating with mine, I hesitated (then) to adopt the term 'civil war'. I stayed in what I thought were the more comfortable zones of cultural conflict and dilemmas of development. In 2008, however, I carried a copy of *The Vertigo of Late Modernity* (Young, 2007) through Dhaka airport and found reading it in

Bangladesh both enlightening and confronting. This is a book of atmosphere, of existential feeling rather than the dry analysis of quantitative criminology. Young's confidence in the previous categories of social science is replaced by vertigo, a 'sense of insecurity, of insubstantiality, and of uncertainty, a whiff of chaos and a fear of falling' (2007: 12). Immersed in the tradition of (leftish) social theory (Durkheim through Merton contra Parsons), Young's world was predominately European, latterly transatlantic, but he was self-critical enough, even if he also realized he could not overcome it, to understand in vague form that he, like others such as David Garland, used phrases such as 'modern society' or late modernity as if this category, this trope, was universal when the referent was actually focused on Western society and, moreover, did not actively position Western society in global networks. Young understood that something was fundamentally limiting with the way in which he had been categorizing (as, for example, he now sought to 'blur the binary vision' which gave him the separation of exclusivity and inclusivity in *The Exclusive Society*, 1999) which culminated in his *The Criminological Imagination* (2011). Young therein recognized that 'the blurred, devious and ironic nature of social reality' threatened the 'taken-for-granted world of everyday life', and that 'critical' meant in part to question 'the solidarity of the social world' (2011: 224). Yet having slightly prised open but not escaped the clutches of methodological nationalism (i.e. to treat the analysis of one's society, nation, the located visage as if it was 'society', as if it was the 'contemporary' rather than an exemplar of the contemporary), he could not move to the global. That is not to say that it is not possible for criminology to confront globalism fully – indeed, criminology calls out for global positioning – but there is no sustained criminology as yet, although some make concerted efforts to clear a path that takes seriously the need to see globally (see, e.g., Aas, 2007/2013; White on developing eco-global criminology, 2011). In *The Vertigo of (Partial) Late Modernity*, Young argued that expressive criminality in 'contemporary society' was linked to a sense of 'bulimia' or constant hunger, where one consumes but is not satisfied and often discards the consumed items in search for greater satisfaction ('it absorbs and it rejects'). Hooked on consumerism, worshipping success, money, wealth and status, but at the same time 'systematically excluded from its realization', offenders suffer a new form of Mertonian frustration. Young half turned to the cultural but felt the need to locate the dilemmas (contradictions) of this exclusion and frustration in terms of 'economic circuits', but nowhere did he touch on the crucial element in the economic circuit: Western debt, government deficit and the distant 'other', beyond the gaze, that carries out the production. Bangladesh contains many who produce the goods for the Western turbo-consumption, but for the country's workers there was no place in Young's vision: in 2007 the minimum wage in the garment industry was around 2,600 taka a month (i.e. US$ 30). Here, often in near-desperate conditions, the objects for Western consumerism are produced. Young saw that 'it is...a consumerism which evokes self-realisation and happiness, but which all too frequently conveys a feeling of hollowness...where commodities incessantly beguile and disappoint. Even the real thing seems a fake' (2007: 3). It is little wonder that an object, the production of which the Western consumer knows virtually nothing about, may be 'other' (see 'Unwatchable: is your mobile telephone rape free', the video for the reality behind the mobile telephone, https://www.youtube.com/watch?v=fIePzz_CEuQ), but the feeling of hollowness in the garment worker may be of physical hunger, and thirst, for, having snatched a lunch crouched near the holes in the floor that fulfil the role of toilets, the worker knows not to drink too much because it will be another four hours at least before they are allowed to move from their machines. There, in the areas surrounding Dhaka, the object to be consumed is all too real, the stitching of which has to be perfect or else penalties may befall, wages withheld.

Yet Young relates in a language in which some of the referents can be reclothed, thus:

> Because of ontological insecurity there are repeated attempts to create a secure base. That is, to reassert one's values as moral absolutes, to declare other groups as lacking in value, to draw distinct lines of virtue and vice, to be rigid rather than flexible in one's judgments, to be punitive and excluding rather than permeable and assimilative.
>
> (Young, 1999: 15)

In Bangladesh, ontological insecurity can take many forms, from the pressing need just to survive to the 'cultural war' that Seabrook (2001) identifies, a clash between 'Bengali culture – pluralist, humanist, rooted, with its songs, dance, folklore, drama and literature' and 'the austere discipline of Islam', a destabilization often occurring in the same people. What is Bangladesh? For Benedict Anderson (1983), nations are 'imagined communities'; for Philip Oldenburg (1985), Bangladesh is 'a place insufficiently imaged'. Bangladesh lies between these two quotes. Anderson links the rise of nationalism to technological developments that allowed linguistic forms to flourish over particular geographical spaces and give rise to popular expression (in the novel, song form, poetry): a nation needed its language and the printing press allowed expression to be captured and communicated. The narratives of the founding of Bangladesh are constantly those of 'reacting', of preserving the language, of surviving 'genocide'. Oldenburg hints at a reason for the continual existential fight over the identity of Bangladesh in a lack of positive conceiving of what Bangladesh would be. Born from the partition of India as East Pakistan – part of the Muslim home – the 'liberation' of Bangladesh was more of a reaction to developments dictated by West Pakistan (such as the imposition of Urdu as the state language, in part a movement away from the secular and democratic image that a founding statesman, Jinnah, had given) rather than an articulated vision of what Bangladesh could become; the narratives of its founding stress reactions, of preserving the language, of surviving 'genocide'. The Awami League formed a government in exile which passed a secular 'constitution' and committed the state not to enforcing a particular image of the good but to being tolerant. It is difficult for liberal constitutions to visualize positive freedom (it often looks banal), hence the importance of the story of liberation and resistance as a trope of negative freedom – we did not lie down and acquiesce. This makes the fate of those who collaborated with the Pakistan military, who formed paramilitary groups which aided in the repression and elimination of Hindu and politically active nationalists, and who at first were barred from politics but later under military governments were reinstated and achieved positions of great influence, while undermining Bangladesh's separate identity, a crucial issue.

In recent years, Bangladesh's growth rate of 5–6% had given hope that it would be a player in the global marketplace, growing its garment industry towards a US$ 20 billion a year industry. At the same time this process entailed in part a ruthless transformation of its traditional rural cultural form, and this would intensify. Perhaps as a consequence of this, the reaction of the Islamic groups – a revival of fundamentalist Islam – was defensive, a feeling that Islam might be in danger, that 'development' was producing a Bangladesh of the new cultural insignia of global liberalist economics in which national boundaries are more porous and the Internet overwhelms the mosque as the site of cultural exchange for the emerging middle class. Certainly many non-governmental organizations (NGOs) see their role as empowering women and thus disrupting the traditions of the village (and thereby fighting what Hartmann and Boyce (1983) described in their 'view from a Bangladesh village' as *A Quiet Violence*; and some of the ex-students of the London programme were now as lawyers taking on 'acid-throwing' cases and thus depriving the local Iman of power over law and order). Thus nostalgia for the Pakistani dream strengthens, and if this is a reactionary identity politics, then it also reflects the lack of any utopia, of whatever form, in the political imaginary.

Although I found it difficult to comprehend, I listened to many who expressed a profound belief that Islam was somehow threatened, that globally it was under attack (the response to 9/11 and the invasion of Iraq were certainly factors), that the new urban youth were embracing 'consumerism' and adopting practices that were far from the five pillars of Islam. Vertigo? Certainly we see in play the project of increasing material wellbeing throughout the world by means of extending capital accumulation. The post-Hobbes process of the pursuit of felicity, of freeing up individuals from disorganized and chaotic war to a sociospatial satisfying of human wants, needs and desires – all the time unrealizable and continual – has been in the past Western-centric. The disasters of the Bangladeshi garment industry, fires and building collapses, show some of the cost of this spatially and hierarchically organized consumption. But we are in an ambivalent position because even if the consumerism of the West demands the cheapest prices, which results in the working conditions in the factories around Dhaka, there are other competitors, the Chinese, the Vietnamese and so forth. The fact that the workers come in their hundreds of thousands from the villages of rural Bangladesh illustrates that the tyranny of the village which Marx depicted in the nineteenth century still continues.

Consider Stan Cohen

On 15 December 2012, *The Economist*, under the heading 'Trying war crimes in Bangladesh: The trial of the birth of a nation', wrote about the resignation of the tribunal chairman. This was in the wake of the magazine receiving and publishing extracts of 17 hours of Skype conversations between 28 August and 20 October 2012, and more than 230 e-mails between September 2011 and September 2012, between the chairman and Mr Ziauddin, a Bangladeshi lawyer and academic based in Brussels.[5] The tapes showed that the chairman was under pressure from the government to issue a judgement, and he complained to his friend and confidant that he simply could not produce one in the time required. Ziauddin – described as both an international lawyer and a criminologist – was the director of the Bangladesh Centre for Genocide Studies at the Catholic University of Brussels. He has long campaigned to end what he called 'the ingrained culture of impunity' regarding the 1971 war crimes in Bangladesh and is also known for his advocacy of the Rome Statute of the International Criminal Court because he lobbied Asian governments to sign and ratify the treaty. Ziauddin had talked with the chairman for at least 20 minutes a day, and had advised on and co-drafted much of the courts (and the prosecution's) documents. He strongly advised the chairman about the need to go slower and to draft judgements carefully as the world would be watching.

It is little surprise then that in delivering its first judgement on 26 January 2013, the second of the war crimes tribunals of Bangladesh quoted Stan Cohen from *States of Denial*: 'after generations of denials, lies, cover-ups and evasions, there is a powerful, almost obsessive, desire to know exactly what happened'.[6] Attempting to place themselves in a global metanarrative, the judgement continued: 'In Bangladesh, the efforts initiated under a lawful legislation to prosecute, try and punish the perpetrators of crimes committed in violation of customary international law is an indicia of valid and courageous endeavour to come out from the culture of impunity.'

This was to align the court with a recognizable ethical imperative. As Anthony Mascarenhas, the journalist whose stark headline in the *Sunday Times* in 1971 – 'Genocide' – brought the events of the West Pakistan military's repression to the West's attention, ended his account of Bangladesh's first 15 years in 1986, 'Machination and murder had been the curse of Bangladesh – its legacy of blood. It will not end until public accountability and the sequence of crime and punishment is firmly established' (1986: 183). The tribunal was the result of a diverse set of campaigns over many years, but part of its account is deeply problematic. Having stressed the importance of 'truth in itself', Cohen later on the same page drew a distinction between knowledge and acknowledgment, where acknowledgement is what happens to knowledge when it becomes officially sanctioned and enters the public discourse. The tribunal presented a certain – officially sanctioned – narrative as 'historical background':

> Atrocious and horrendous crimes were committed during the nine month- long war of liberation, which resulted in the birth of Bangladesh, an independent state. Some three million people were killed, nearly a quarter million women were raped and over 10 million people were forced to flee to India to escape brutal

persecution at home, during the nine-month battle and struggle of Bangali nation. The perpetrators of the crimes could not be brought to book, and this left a deep wound on the country's political psyche and the whole nation. The impunity they enjoyed held back political stability, saw the ascent of militancy, and destroyed the nation's Constitution (para. 4).

The judgement then immediately references two international scholars. A well-known researcher on genocide, R.J. Rummel, in his book *Statistics of Democide: Genocide and Mass Murder since 1900*, states:

> In East Pakistan [General Agha Mohammed Yahya Khan and his top generals] also planned to murder its Bengali intellectual, cultural, and political elite. They also planned to indiscriminately murder hundreds of thousands of its Hindus and drive the rest into India. And they planned to destroy its economic base to insure that it would be subordinate to West Pakistan for at least a generation to come.

Women were tortured, raped and killed. With the help of its local collaborators, the Pakistan military kept numerous Bengali women as sex slaves inside their camps and cantonments. Susan Brownmiller, who conducted a detailed study, has estimated the number of raped women at over 400,000 (http://bangladeshwatchdog1.wordpress.com/razakars/%5D; websource in original, para. 5).

Rummel is an author who has done valiant work in attempting to establish the huge, dark figure of state-sponsored crime. Brownmiller's famous 1975 text was rightly characterized as an 'overwhelming indictment' of the prevalence of rape and a critical metanarrative of the operation of the 'house of the Father's' where women were protected from male aggression at the cost of reinforcing a system that perceived them largely as property. Her actual estimate was between 200,000 and 400,000. Both writers worked with secondary accounts, accepting sources that are now open to contestation.

The judgement rightly noted that Bangladesh is a signatory to various international conventions, and it stated that the proceedings were in line with those commitments; it explained 'the degree of fairness' that is embedded in the act, and the Rules of Procedure formulated by the tribunals under the powers conferred in Section 22 of the principal act 'are to be assessed with reference to the national needs such as, the long denial of justice to the victims of the atrocities committed during 1971 independence war and the nation as a whole'. It explained that it would accept as given a history of 'common knowledge'. At Para 10 it set the scene for the question: Why try collaborators?

> In the War of Liberation ... all people of East Pakistan wholeheart-edly supported and participated in the call to free Bangladesh but a small number of Bengalis, Biharis, other pro-Pakistanis, as well as members of a number of different religion-based political parties, particularly Jamat E Islami (JEI) and its student wing Islami Chatra Sangha (ICS) joined and/or collaborated with the Pakistan military to actively oppose the creation of independent Bangladesh and most of them committed and facilitated the commission of atrocities in vio-lation of customary international law in the territory of Bangladesh. As a result, 3 million (thirty lac) people were killed, near about quar-ter million women were raped, about 10 million (one crore) people deported to India ...

Again the figures of 3 million killed and of 250,000 raped are taken not as assertions but as common knowledge. It is undeniable that large-scale atrocities occurred in 1971 and that the vast majority of them were caused by the anti-liberation forces, whether the West Pakistan military or the paramilitary supporters whom they set up. The court accepts as evidence a range of material (newspaper reports, personal accounts and accounts of books), and the court (and the prosecution) accepts these accounts as setting the context for the supposed actions of those who are accused. However, the evidential basis for partic-ular claims as to the accused and evidence of particular events are weak. Many of the accused have been convicted on very weak evi-dence. This is, however, as Nancy Combs demonstrates in her study of the evidence that is used to convict in other 'international criminal' courts, not unusual (Combs, 2010, who entitles her work *Fact-Finding without Facts*). But here the institutional power of the court writes history – an institutionally accepted history that is reinforced by the power to hold in contempt of court those who question it – as David Bergman, a journalist who runs a blog about the trials and who had posted an intelligent and searching piece on the history of the numbers quoted, found out (he was convicted of 'contempt of court' in 2014). When the defence team attempted to introduce as academic writing (to counter the piles of accepted texts, newspaper cuttings and 'mem-oirs' that were introduced by the prosecution) the revisionist history of Sarmila Bose (2001), they were blocked (personal communication,

2013). Cohen suggested that 'Historical skeletons are put in cupboards because of the political need to be innocent of a troubling recognition; they remain hidden because of the political absence of an inquiring mind' (2001: 139). We are here dealing with the converse. The continual reference to the War of Liberation is because of the need for a process of recognition, but the stylization of the process constrains the enquiring mind.

Consider Stuart Hall

In 1978, *Policing the Crisis* was published, analysing the events of the early 1970s in the UK. The British state was depicted as experiencing a 'crisis in hegemony' and an 'exhaustion of consent'. In partial response, a moral panic of the mugger and a crackdown on 'law and order' was created in a general sense of 'authoritarian populism', in which the state sought to retain control of refractory populations by 'orchestrating public opinion' around a threatened eruption of a violent society. As the authors saw it, mainstream Britain feared immigrant 'barbarians at the gate', and for those already let in, the mugger was the 'enemy within'. He signified the arrival of alien values, alien cultures and the disintegration of a mythical English past. Fears and anxieties about other processes were displaced onto black people, and all blacks were potentially criminal.

The cultural picture was of a crisis in hegemony, in political leadership and ideological domination of the society. The ruling elite needed to win support for their policies and ideas from other groups in society to remain in power. They tried to persuade the working class that the authority of the state was being exercised fairly and justly in the interests of all. The path was painted of a journey from an established state, to one suffering a crisis of hegemony, and on through a law-and-order moral panic into an 'exceptional state'.

The Awami League government in 2013 seemed to follow the script perfectly. The target was those who questioned the legitimacy and operation of the trials and of the state, and whether the state could find itself on the narrative of the War of Liberation. But was Bangladesh ever a normal state? Born into the wreckage of the War of Liberation, its first government had a written constitution but no grasp of handling the multiple social and economic problems, assassinations and military governments that followed in 1975, and democracy was not restored until 1990. Since then it has had a politics of sectarian interests, one that is semifeudal and where the spoils of power, rather than ideology, drive political ambition.

And was this a manufactured enemy? Or is the 'other', actual, real and dangerous?[7]

In place of a conclusion: Theoria and the abyss.

In *Crime and Everyday Life* (2002: 3), Marcus Felson relates that the majority of crime lacks drama and is 'not much of a story'. In *A General Theory of Crime* (Gottfredson and Hirschi, 1990), the authors argue that their theory can encompass the reality of cross-cultural differences in crime rates (175) and they state that 'a general theory of crime must be a general theory of the social order' (274). However, they then admit that lack of time and space prevented them from pursuing this issue further and they restrict their view to the official data on crime in the USA, buttressed by victim surveys and self-report studies. As a result they conclude: 'nearly all crimes are mundane, simple, trivial, easy acts aimed at satisfying desires of the moment' (xv). Recently Gottfredson felt compelled to argue that it is necessary to overcome the 'false image...that depicts crime as planned, as crafty, as requiring "tough" offenders' (2011: 39).

So mainstream criminology seeks to render the object of criminological vision mundane and uncomplicated. Against this I have suggested that criminology should escape from its blinkered vision of the everyday and embrace theoria, and incorporate places outside its normal gaze. In Bangladesh, everyday life is drama. It is a place where any difference between crime and politics is blurred, ironic and contingent. The foundation of the everyday is fought over and always in danger of slipping away. The past haunts – is never past. The juxtaposition of the normal and the exception is reversed. For example, the Interpol figures for 1998 (around when I first visited) recorded 3,539 voluntary homicides in a population of about 130 million, while a figure of 1.5 million would be reasonable for the numbers killed or dead as a result of illness caused by the war in the nine months from March to December 1971 out of a population of around 75 million. Thus it would take around 424 years of the 'normal' 1998 homicide rate to equal the nine months of human gestation before birth.

Critical criminology looks at everyday life differently. Hall and others look behind the games of the state and find law-and-order politics that may lead towards our accepting the state claiming powers of exception; Young finds vertigo, contingency and a constant fear of falling; Cohen tells us that justice is divided, that suffering always occurs somewhere else, and he relates how when he looked at the world he felt a sense of guilt, always had the feeling that something was basically wrong, but could not understand why others – most people – could not see the world as he did. Viewing Bangladesh confirms and yet confronts the aura of critical criminology.

Bangladesh is not somewhere else; it is where a remarkable number of the clothes that we wear are manufactured. When its garment factories burn or collapse, it is partly due to our demand for cheap and constantly changeable consumer items. Liberal voices look at the war crimes trials and find 'another kind of crime' ('Justice in Bangladesh: Another kind of crime', *The Economist*, 23 March 2013). They also warn of the abyss ('Bangladesh war crimes tribunal in the dock', Al Jazeera, 15 July 2013; 'Flawed war crimes trials push Bangladesh to the edge', *Financial Times*, 2 October 2013). For those hoping for an international rule of law, this is 'justice denied', and while all accept that 'war crimes trials are a defining moment' for Bangladesh, it has sullied 'its judicial and political systems', the trials have poisoned 'the well from which Bangladesh will one day want to drink' (*The Economist*, 23 March 2013). However, many Bangladeshis remember that few outsiders supported the nation's creation (the USA in particular backed Pakistan) and think that such criticism is paid for by Jaamat-e-Islami's foreign money (personal communication from victims; see also Khan, 2014).

Social justice, social solidarity and democracy – these are the tropes of critical criminology, but each move in Bangladesh shows their contested and multiple sides. The power to define who is a criminal and who is a legitimate political opponent is the major 'othering' of Bangladesh. Bangladesh was constituted in the 'mystical authority' (Derrida, 1992) of the violence of the War of Liberation. Today its government uses the 'power of law' to state and restate its version of the metanarrative of Bangladesh. To enter into this as a committed observer is to emphasize with those striving for justice and social solidarity in a continual cycle of that violence, to look into the abyss knowing that the abyss is also looking at you.

Notes

1. There are numerous reports that one could quote, but as an example, *The Economist*, 17 September 2013 ('The gallows, not jail, had always seemed like the more likely destination for Abdul Quader Mollah') reported that an opinion poll of April 2013 showed that 'though nearly two thirds of respondents said the trials were "unfair" or "very unfair", a whopping 86% wanted them to proceed regardless'. However, it warned of the prospect of 'a political backlash and perhaps even the rehabilitation of the BNP's [Bangladesh National

Party's] Islamic allies; a less secular Bangladesh; and millions of Awami League supporters incensed by their government's utter failure to deliver credible war-crimes trials'.

2. Under the heading 'Another beating [for democracy]: Sheikh Hasina plans to hang on to office after an electoral farce', *The Economist* (9 January 2014, 10: 33) stated that 'It is becoming hard to know whether Sheikh Hasina, Bangladesh's prime minister, is a cynically good actress or cut off from political reality.' Addressing the election outcome, Prime Minister Sheikh Hasina had blamed the main opposition force, the BNP, and announced that she would not enter talks about a new election unless the opposition first renounced violence: 'Today, democracy is tainted by the blood of innocent people and soaked by the tears of burned people, who have fallen victim to the violent political program that is hitting the nation's conscience.' She added she had ordered the army to 'curb any post-poll terrorism and violence with iron hands' ('Bangladesh's governing party wins vote amid unrest', *New York Times*, 6 January 2014).

3. Bangladesh is now being enclosed. India is constructing the Indo-Bangladeshi barrier, a 3,406 km (2,116 mile) fence of barbed wire and concrete, just under 3 m high, As of March 2011, 2735 km of fencing was completed. Assam shares a 263 km border with Bangladesh, out of which 143.9 km is land and 119.1 km is riverine. As of November 2011 some 221.56 km of fencing was completed. For the effects of the fence, see Hussain (2013). For some, this is turning all of Bangladesh into a prison.

4. For the reality of the sex trade, see Brown (2000). Dhaka now has numerous coffee bars that cater for the growing middle class, including one opposite the (unnamed) hotel above.

5. The Bangladeshi newspaper *Amar Desh* also received the conversations, and it published a report on 9 December, followed by the transcripts in full. On 13 December a court injunction banned Bangladeshi newspapers from publishing the materials, at which time *Amar Desh* stopped further publication. The editor later faced contempt and sedition charges.

6. At para. 47. The reference to Stan Cohen was to *States of denial*: Knowing about Atrocities and Suffering (2001: 225). The judgement was in the case of Abul Kalam Azad, who was charged with genocide, rape, abduction, confinement and torture, and was tried in absentia after having fled the country.

7. Consider, for example: 'In this subcontinent, if we [Bangladesh] and Pakistan unite, we won't have to care about Malaun [Hindu], atheist and Nasara [Christian]. We will name this Islamic country as Islamic Republic of Banglastan or join with Pakistan where only Muslims will live' (Facebook post of Basher Kella [run by Jamaat-Shibir activists, i.e. an Islamic group] posted 3 March 2013. The *Daily Star* reported 100,000-plus followers and that a posting of 28 February instructed the Jamaat-Shibir men on tactics for the next day's countrywide hartal (violent strike). Among the 10-point suggestion were uprooting rail lines, snapping road communications between Dhaka and other districts by barricading, setting fire to homes of all lawmakers and ministers, spreading a smear campaign against police, and attacking law enforcers and journalists. Consider also: 'Don't push the country into a civil war by delivering one-sided verdicts against our leaders. If anything happens against Quader Mollah, every house will be on fire' (Jamaat acting secretary general

Rafiqul Islam Khan, in a press release declaring a 'hartal' (nationwide strike) on 5 February 2013, awaiting the war crimes trial verdict on Jamaat's assistant secretary general, Quader Mollah).

Bibliography

Aas, K.F. ([2007] 2013) *Globalization and Crime*, 2nd edn. London: Sage.

Ahmed, I. (2009) *Historicising 1971 Genocide: State versus Person*. Dhaka: The University Press Ltd.

American Heritage Dictionary of the English Language (2009) 4th edn. Boston, MA: Houghton Mifflin.

Anderson, B. ([1983] 1991) *Imagined Communities: Reflections on the Origin and Spread of Nationalism*, revised edn. Verso: London.

Ashraf, A. (2013) 'Bloodthirsty non-violence: the paradox of Shahbag Protests, First Post at http://www.firstpost.com/world/bloodthirsty-non-violence-the-paradox-of-shahbag-protests-651599.html

Beccaria, C. (1788) *Von Verbrechen und Strafen* [On Crimes and Punishment]. Breslau: Johann Friedrich Korn.

Bose, S. (2001) *Dead Reckoning: Memories of the 1971 Bangladesh War*. London: C. Hurst & Company.

Brown, L. (2000) *Sex Slaves: The Trafficking of Women in Asia*. London: Virgo Press.

Cohen, S. (1996) 'Crime and politics: Spot the difference', *The British Journal of Sociology*, 47(1): 1–21.

Cohen, S. (2001) *States of Denial: Knowing about Atrocities and Suffering*. Cambridge: Polity.

Collins Thesaurus of the English Language (2002) 2nd edn. New York: HarperCollins.

Combs, N.A. (2010) *Fact-Finding without Facts: The Uncertain Evidentiary Foundations of International Criminal Convictions*. Cambridge: Cambridge University Press.

Derrida, J. (1992) 'Force of law: "The Mystical Foundation of Authority"'. In Cornell, D. and Rosenfeld, M. (eds) *Deconstruction and the Possibility of Justice*. New York: Routledge.

Felson, M. (2002) *Crime and Everyday Life*, 3rd edn. Thousand Oaks. CA: Sage.

Gottfredson, M. (2011) 'Some advantages of a crime-free criminology'. In Bosworth, Mary and Hoyle, Carolyn (eds) *What Is Criminology?* Oxford: Oxford University Press.

Gottfredson, M. and Hirschi, T. (1990) *A General Theory of Crime*. Stanford: Stanford University Press.

Hall, S., Crticher, C., Jefferson, T., Clarke, J. and Roberts, B. (1978) *Policing the Crisis: Mugging, the State and Law and Order*. London: Macmillan.

Hussain, D. (2013) *Boundaries Undermined: The Ruins of Progress on the Bangladesh-India Border*. London: C. Hurst & Company.

Khan, M. (2014) 'International war crimes tribunal: A performance review', *Dhaka: The Daily Star*, 1 January 2014.

Mannheim, H. (1965) Comparative Criminology. London: Houghton Mifflin.

Mascarenhas, A. (1986) *Bangladesh: A Legacy of Blood*. London: Hodder and Stoughton.

Morrison, W. (1995) *Theoretical Criminology: From Modernity to Post-Modernism.* London: Cavendish.

Morrison, W. (2006) *Criminology, Civilisation and the New World Order.* Abingdon: Routledge/Cavendish.

Morrison, W. (2013) 'Bangladesh, 1971, war crimes trials and control of the narrative: The state or collaborative enterprise?' *Revista Critica Penal y Poder* (Special issue: Redefining the Criminal Matter: State Crimes, Mass Atrocities and Social Harm) 5: 338–357.

Nietzsche, F. (1961) *Thus Spoke Zarathustra*, trans. R.J. Hollingdale. New York: Penguin.

Oldenburg, P. (1985). '"A Place Insufficiently Imagined": Language, Belief, and the Pakistan Crisis of 1971'. *The Journal of Asian Studies*, Vol. 44, No. 4: 711–733.

White, R. (2011) *Transnational Environmental Crime: Toward an Eco-global Criminology.* Abingdon, Oxon: Routledge.

Young, J. (1999) *The Exclusive Society.* London: Macmillan.

Young, J. (2007) *The Vertigio of Late Modernity.* London: Sage.

Young, J. (2011) *The Criminological Imagination.* Cambridge: Polity.

4

A Critical Gaze on Environmental Victimization

Lorenzo Natali

> When elephants fight, it is the grass that is crushed.
>
> (African proverb)

Introduction

In an early contribution, Nigel South (1998) poses some of the following questions: Why do we feel the need for a green criminology? On which already extant works could it be built and what are the theoretical questions that can be asked? More than 15 years later, these questions still prove useful and effective in approaching the multiple contributions of green criminology that have emerged in the meantime. In fact, dealing with environmental crimes continues to imply a necessary confrontation with some extremely complex issues.

This chapter examines processes of environmental victimization and suggests some theoretical frameworks that are useful in contributing to the debate about green criminology, highlighting the importance of empirical research on environmental victims. Next, it proposes a processual model in order to analyse the various steps through which an environmental problem may turn into radical action and social change. Finally, it identifies some possible directions for green criminology to approach environmental victimization and suggests that a re-evaluation of fraternity (*fraternité*) could act as an important transforming force to better tune and synchronize our actions towards the environmental and social worlds, creating more inclusive socioenvironmental relations. As this chapter demonstrates, critical criminology may continue to help green criminology in analysing the power dynamics that contribute to the social reality of environmental crime. This is a fundamental issue if we want to enhance the criminological imagination (White, 2003) in order to contribute to change the way we think and respond to environmental crimes.

An environmental victimology approach

In a recent contribution, Michael Lynch (2013: 45–48) compares the different levels of victimization that are created by environmental crimes with those that are attributed to 'street crimes'. His analysis shows the large number of victims that are missed by traditional criminological approaches, which are still little inclined to take into account environmental crimes and their serious consequences (Lynch et al., 2013: 998). In a way not dissimilar from what happens to the victims of white-collar crimes, environmental victims also often remain hidden in the shadows (White, 2011: 109; Davies et al., 2014). Technological and scientific developments have in fact introduced into our societies totally new vulnerabilities that translate themselves into serious but elusive environmental risks and harms. In this sense, every technology produces specific and unpredictable 'accidents' (Halsey, 2013). In particular, Matthew Greife and Paul Stretesky (2013: 151–152) highlight how the increase in the levels of pollution is a result of Western capital that was invested in new technologies in order to increase production. They also point out that one of the most dangerous aspects of this process is the fact that the visibility of the relationship between production and environmental disorganization lessens over time, together with the obfuscation of both the environmental harms caused and the consequences for the victims.

While the growing field of green criminology has devoted much attention to the study of environmental crimes, the processes of victimization still remain little observed (Bisschop and Vande Walle, 2013: 34–35; Hall, 2013: 218). Thinking about this issue was initiated by Christopher Williams (1996, 1998), according to whom environmental victimology could be placed within a theoretical frame known as 'critical victimology' (Mawby and Walklate, 2000), which focuses on harms to the environment and to peoples' health – harms that may stem from acts and omissions that are not proscribed by law. In this sense, a radical green victimology approach complements the broader definition of environmental crime.

Environmental victimization poses a series of new questions that the systems of criminal justice find themselves unprepared to face (Hall, 2013: 219–220). First, the harms suffered can involve an extended group or even a community of victims, sometimes representing rival interests.

Second, the perpetrators are often corporations or states (White, 2011: 103–104; 2013) – and here we see the importance of developing a notion of 'crime' that encompasses those 'lawful, but awful' acts and omissions (Passas, 2005; see also Hall, 2013: 221; Hillyard et al., 2004; Lynch et al., 2013: 999). Finally, the causality nexus is extremely complex to reconstruct, sometimes leading to a consideration of environmental crimes as 'crimes without victims' (see Bisschop and Vande Walle, 2013: 40).

The relevant scientific literature also clearly shows how the difficulties that are encountered in establishing a causal relationship may offer an easy way of escape for the perpetrators (Williams, 1996). Systematic use of techniques of denial of harm and responsibility further undermine efforts to create causal connections between offenders and victims. In fact, the various strategies of neutralization of responsibility on the part of corporations or the state include denying the problem (Cohen, 2001); neglecting to put into perspective what is seen as damaging (e.g. the long-term benefits); and reproaching, blaming, dividing and confusing the victims (Williams, 1996). For all of these reasons it is important to explore the nature of victimization as an active social process which implies relationships of power, control and resistance (White, 2011: 106; Natali, 2013b).

The need for empirical research about environmental victimization: Digging the path

Social and cultural perspectives are crucial in order to problematize and determine what constitutes environmental victimization (Hall, 2013: 225–226). In many cases, to understand the different narratives that orbit around a case of environmental crime (broadly understood), it is necessary to understand the perceptions of that harm from the inside, starting from the symbolic and cultural perspectives that are expressed by the social actors affected (Natali, 2010; see generally Brisman and South, 2013, 2014). In other words, we still do not know enough about the way in which the human victims of an environmental crime see and interpret or ascribe meaning to the situation in which they live (White, 2011: 121; see also Brisman and South, 2013, 2014). The number of studies available is not sufficient to provide detailed data on the lives of people who live in polluted areas, describing from their point of view what they know of, think about and feel towards the reality in which they are living (Natali, 2010).[1] Such an orientation raises the following questions: How do people live and make sense of their experiences in polluted places? What is the relationship between the knowledge of the risks that are present in a contaminated environment, the experiences of environmental suffering and injustice lived by the inhabitants, and their responses to these threats and harms experienced in the first person? And what does, or can, one expect from the system of justice (Bisschop and Vande Walle, 2013: 49; see also Brisman, 2013; White, 2013)?

Symbolic interactionist Howard Becker refers to folk art not as products of rural remnants of customs that were popular in times past but as activities that are performed outside professional art worlds – works 'done by ordinary people in the course of their ordinary lives' (see Becker, 1982: 246). In this sense, the vocabularies and narratives that circulate among the inhabitants of an area about an environmental issue with criminological relevance represent what could be defined by the expression 'folk green criminology': knowledge and insights – often ignored by the academic worlds of criminology – that are held and created by the 'common people' during their daily experience of the reality of an environmental crime affecting the area in which they live.[2] Empirical research on the inhabitants of highly polluted places can reveal the multiple dialogues that each person continually weaves within themselves – in Herbert Mead's (1934) sense of 'internal conversation' – and with others (Athens, 1994; Natali, 2013b, 2015; see also Cianchi, 2013: 55–56). These dialogues reflect their decision to act individually or collectively (or to abstain from doing so) in response to their environmental victimization. Moreover, social and cultural perspectives on the processes of environmental victimization prove to be extremely important in showing that experiences of environmental suffering are not simply individual: they are actively created, starting from the position that the residents as a group occupy in the wider social macrocosmos (Auyero and Swistun, 2009: 159). In the process of learning that characterizes our way of communicating according to certain schemes and certain metaphors (Kane, 2004: 316), favouring some words and 'universes of vocabulary' rather than others, a significant role is certainly played by those social actors who have the power that is necessary to structure certain discursive universes of the public sphere. Thus green criminologists must confront the marginalization of 'voices from below', recognizing them as 'valid forms and producers of knowledge' (Mol, 2013: 251). Taking into account this form of knowledge means rethinking the hierarchy of knowledge itself, critically analysing the purported clear division between knowledge and non-knowledge,[3] between what can be recognized as 'scientific' – and for this same reason 'real' – and what instead remains at the margins of knowledge.

When undertaking these empirical investigations, it is essential to question simplistic concepts of how the victims relate to the 'uncomfortable truth' of pollution, noting that they themselves often do not always agree on the definition and interpretation of that reality. Contrary to what is identified in the predominant scientific literature, social experiences of environmental suffering are full of doubts, disagreements, suspicion, fears and hopes (Auyero and Swistun, 2009: 4–5). In particular, focusing on the mechanisms of denial (Cohen, 2001; Pulcini, 2013) will also help to understand silence, apathy and a range of other possible responses by those who witness daily the destruction of the environment that they inhabit (Williams, 1996; Brisman, 2012; Natali, 2013b; White and Heckenberg, 2014: 186–192). In fact, in the face of the dramatic environmental transformations that shake their internal and social 'geographies',[4] sometimes victims learn to accept irreparably altered landscapes, sometimes they simply 'delete' them, as one does with an illness or death (Settis, 2010: 73–74).

As philosopher Elena Pulcini (2013) highlights, however, if denial is useful in explaining the failure to perceive a given danger and the emotional numbness linked to this ignorance, another mechanism that may help to explain the complex emotional responses that the social actors exhibit is self-deception (Pulcini, 2013: 121). In the environmental contexts considered, it is not only a question of defending oneself emotionally from unbearable events (denial) but also of persisting with actions that allow people to legitimate and satisfy their actual desires. The flows of desires can continue to be lived as flows of pleasure only thanks to a process of immunization that anaesthetizes fear and nips in the bud any potential action for change (Pulcini, 2013).[5]

A processual model for radical action and social change

In *Ecopopulism*, sociologist Andrew Szasz (1994) considers victimization to be an 'active social process'. Referring in particular to the issue of industrial toxic waste, he asks in what way an environmental issue, initially of little relevance, can gain traction, and how various social perceptions of an environmental problem spur radical actions.

While it is difficult to proceed from the 'mute physical fact of damage' (Szasz, 1994: 30) that constitutes environmental harm to an issue of social and political salience, it is possible to indicate some steps that are involved in this transformative process. First, some social actor must notice a series of effects and suspect a common origin. Second, those who are suffering harm must convince a significant part of society of the real existence and seriousness of the perceived damage and persuade them that the harm is actually being caused by a particular industrial activity. This is really a battle in defining the situation. Finally, the problem must be expressed in political terms and defined as a request for legislative, judicial or enforcement action by the state.

As one can imagine, each of these procedural steps is unlikely to be straightforward. For example, it may take years for some environmental harms to become known or obvious (Natali, 2013a). In addition, even if there is little lag time between an act or omission that results in a harm and some sort of manifestation of that harm, causal attribution can be extremely complex, such as trying to link exposure to chemical agents and the illnesses contracted by workers and inhabitants of polluted areas (Szasz, 1994: 30–31). Finally, powerful actors can use their influence to delay and/or weaken the efficacy of the law (Tombs, 2013: 272; see also Stretesky et al., 2013).

As such, it becomes crucial to facilitate the various steps towards social and political recognition, and the visualization of any harm that is connected to industrial activity. This process can be aided by casting light on that 'twilight state' where environmental harms arise as happenings that don't yet exist in the social and discursive sphere (Szasz, 1994: 31). It will, however, never be a 'full' light because of the various operations of denial (Cohen, 2001) and corporate 'green-washing' that are endemic in the environmental field (see, e.g., Lynch and Stretesky, 2003; see also Walters, 2010: 315; Ellefsen et al., 2012).

Joining associations and movements can provide an individual with a chance to transform one's private experience of environmental suffering into a public experience of participation – a path that is 'made of discovery, learning, awareness and demand for recognition of one's rights' (Altopiedi, 2011: 107). Such interactions with similarly situated others can help one to understand the harm as injustice, whereby one can learn 'new criteria of valuation of one's own and other's behaviour, redefine rights, duties and responsibilities' (Altopiedi 2011: 118; see also Natali, 2010). In particular:

> The problematization of the normal, of the obvious, carries with it a moral valuation, the ascription of blame. The accusation of injustice, of violence, produces a different awareness and knowledge, what before was considered *normal*, is then defined unjust and oppressive.
> (Altopiedi, 2011: 115)

This process becomes even more important when we consider that state complicity in the routine and systematic production of corporate damage impedes the recognition of corporate crimes as 'real' crimes (Tombs, 2013: 276; Altopiedi, 2011). The public can be anesthetized against such damage, to see and not to see at the same time, which is a most pernicious effect (Tombs, 2013: 284). Thus the relationship between visibility, recognition and active resistance becomes extremely vital: in this sense, it is crucial to recognize the forms of 'cultural' violence that are used by the powerful and by hegemonic culture that defines and redefines reality and that obscures the (individual and structural) violence that is perpetrated and the harms caused (Cottino, 2005: 12–14, 59–61).

Beginning to reconstruct and redescribe these crimes as real and violent may be possible through a 'process of *unveiling* of the mechanisms of legitimization, justification and obfuscation' that characterize them (Altopiedi, 2011: 116). It seems useful to recall here the Gramscian notion of hegemony in the sense of cultural direction, differentiating it from the concept of domination as direction based on pure coercive force (see Angioni, 2011: 175). Where one includes within a wide notion of power both the idea of hegemony and that of domination, the understanding and the analysis of contemporary worlds become even more penetrating and able to take into account the multiple vertical and horizontal dimensions of power itself (Angioni, 2011: 182–191). To follow this path it will be necessary to develop a new reflexivity about phenomena that are not yet adequately explored, such as environmental victimization (Bisschop and Vande Walle, 2013: 48). Green criminology will play an important role in that respect.

Vulnerability, technology, law and fraternity: Some possible directions for green criminology

To bring about an adequate re-evaluation of the horizons that delineate the most salient contours of environmental victimization, it is necessary to reimagine our borders – both disciplinary and non-disciplinary – in such a way as to make room within them for the dimension of vulnerability. Judith Butler (2013) analyses the socioinstitutional processes and structures that legitimate and enhance or, as the case may be, devalue and render unworthy of protection, certain forms of life. This is also true with regard to the multiple dimensions of environmental victimization. As Butler (2013: 22, 45) reminds us, human and non-human beings cannot 'survive and go on independently from an environment that sustains [them]'. To affirm that means then to recognize the relational dimensions of (in)justice and our vulnerability before others and the institutions themselves.

In this sense, rather than deny vulnerability or exploit it, it would seem appropriate to continue to 'produce the conditions in which vulnerability and interdependence become liveable' (Butler, 2013: 62). In what way can criminology also contribute to produce these conditions? In the field of green criminology, this might happen by placing at the centre of criminological reflections the constant interrelation between the various forms of power, justice and harm that manifest themselves in environmental crimes (see Walters et al., 2013).

Technology plays an important role in creating the conditions of vulnerability that are present in the micro- and macrotragedies that involve the ecosystem (Baratta, 2006: 62–63). We know, however, that technology is double-edged: 'not one good technology that cures and one bad technology that makes ill, but the same technology cures us, making us ill, saves us killing us; the same and at the same time' (Resta, 2008: 107–108). Technology appears, thus, as a 'poisoned gift'. It partakes in the basic ambivalence that Plato associates with the idea of *pharmakon*: it is illness and remedy at the same time. From such a perspective, technology always suggests a peculiar vision of things and reality (Sini, 2009: 24–25). Therefore it is not 'good', or 'bad', or 'neutral', and its manifestations are neither utopian nor dystopian: they manifest what we are (Castells, 2001; see also White, 2011: 90–91). If it is true that it is important to understand not only human inventions but also the impact that they have on human beings in relation to a feedback effect (Sini, 2009: 13), particular attention should be paid to the consideration that technological tools trigger, as any innovation does, 'complex and unpredictable twists among practices of life and thinking' (Sini, 2009: 36; see also Halsey, 2013; Kane and Brisman, 2013). Only if we feel responsible for what we do and for what happens around us will it be possible to control and to guide this unprecedented technological creativity (Castells, 2001).

On the other hand, in our industrialized societies, it is not conceivable, except in the shape of an abstract utopia, to think of really 'going back to nature' – of suddenly and at once stopping our system of production and consumption (Langer, 1961/1995: 166). The world has changed too much to allow us to turn back time and return to a previous era. To exclude the possibility of a return to a romanticized pre-industrial world does not, however, mean to accept uncritically the notion and the usage of time that prevails in the societies in which we live. Again, Eligio Resta (2008: 86) highlights how it is really the ambivalence that is embedded in the concept of *pharmakon* – remedy and poison at the same time – that ties law and technology. If the code of technology is 'we can do all that we can do', the law should then be so strong as to add a question mark to this statement, 'putting out of sync' the time imposed by technology and thereby introducing a 'setback' that interrupts its sagittal and monologic character (Resta, 2000).

However, a view of law as a punitive instrument is not enough to this end. To break this temporal perspective (Natali, 2013a) also means to be open to the dimension of *fraternité* (Resta, 2004; Ceretti and Cornelli, 2013: 206–207). Nowadays, *fraternité* presents itself with all of its paradoxes but also with the same transformative strength. Some scholars have highlighted its ambiguous character: if filled with fear of the other, *fraternité* can become once more the premise for collective selfishness that translates, in the political field, into regressive instances of the democratic spirit. If, on the other hand, it is combined with a political project of broad views, it can become its principle of orientation (Ceretti and Cornelli, 2013: 206). It is with this latter orientation that it is possible to create open and inclusive social relations, beginning from the awareness of sharing a common condition of vulnerability on the social, economic and environmental level. From this philosophical perspective, a fraternal law may be described as similar to the rhythmic formula that is known in the world of music as *contrattempo*, serving to shift our attention to an upcoming transition (Resta, 2000; Resta, 2008: 110; Vasta, 2010). With this suggestion in mind, it seems plausible that the re-evaluation of *fraternité* could become the transforming force that renders our conduct towards our environment ecologically and socially tenable, thereby reducing environmental victimization through the gradual development of an ethical protection of the environment and of all those within it – men, women, children and non-human animals, as well as ecosystems (see also Zaffaroni, 2012: 144). An idea of *fraternité* extended to the whole of nature and its individual parts – a 'concrete utopia' (Bloch, 1959/1986) – could be the start of a new theoretical basis for legislation that protects the environment. As Eugenio R. Zaffaroni (2012: 127) pointedly remarks concerning the ecological question,

> Only substituting the knowledge of the *dominus* with that of the *frater* can we regain human dignity…Constitutional ecology, within the vision derived from our original cultures, rather than denying human dignity, will re-set it on its path now lost because of the desire of domination and infinite hoarding of things.
>
> (Author's translation)

Conclusion

The importance of developing green criminology and environmental victimology, and of recognizing their critical roots, is evident when we consider that many cases of routine, systematic and ongoing environmental harm continue to occur without being effectively challenged. More importantly, these forms of harm are, as in the case of pollution, often 'supported by the State that makes it possible and even normal, to the point that the states should be seen as complicit in its production and its violent effects' (Tombs, 2013: 267). Social and environmental harms affect real people – flesh and blood people (Baratta, quoted in Fiandaca, 2013: 108) – and non-human animals. Social and environmental harms are real even when they concern apparently evanescent subjects, such as the environment or 'collective' victims.

In contemporary societies, the image of oneself as a 'spectator' of reality often manages to obscure that of potential (environmental) victim into which each of us can turn, anywhere on the planet (see Pulcini, 2013: 150). So how can we learn to see and make visible the latent possibilities that already live in our present and use them as starting points to build alternative environmental scenarios? According to Pulcini (2013: 139), it is necessary first and foremost to reactivate a 'feeling' that is capable of measuring up to the enormity of the challenges and transformations occurring, and, more generally, able to renew a 'contact' with the world. Only by increasing the volume of our imagination and of our feelings can we still hope to measure up to our actions and to the environmental changes that are caused by them, winning back, in this way, the role of possible 'makers of our future' (Pulcini, 2013: 147). Thus imagination really becomes transformative.

The incapacity of criminological knowledge to ask the decisive question 'What distinguishes and at the same time unifies the social with the natural sphere?' comes mainly from the difficulty of approaching the dichotomies between persons and things, between technology and society, between culture and nature that ground the conceptual structure of traditional criminology. In this sense, to develop a 'criminology of hybrids' (Brown, 2006; Natali, 2013a) that is able to weave these polarities, not as binary contradictions but as a complexity, becomes extremely important in order to understand and formulate preventive measures that are adequate for the new hybrid forms of sociality and domination that produce environmental disasters (Natali, 2013a; see also Kane and Brisman, 2013: 102; see Larsen, 2012). From this perspective, questioning the dichotomies that are often produced by abstract causal models is a difficult but necessary task if we want to develop a complex and 'radically thick' understanding (Lasslett, 2013: 100–101) of the harmful consequences of environmental crimes, in their specific regional contexts, and of the multiple forms of environmental victimization. Among the hybrid events that 'knock on the door' of criminology, environmental risks and harms, together with the dimensions of injustice and responsibility that are tied to them, are some of the most significant of our time (Natali, 2013a). To invite these events to enter the criminological building necessarily implies a rethinking of the structures and of the architectural lines, of the spaces and of the innumerable 'rooms' that, within our field, accommodate various disciplines and different kinds of knowledge, sometimes in communication, sometimes totally deaf to each other. By giving reflexive space to the ideas of vulnerability, technology and *fraternité* it will be possible to extend the already relevant theoretical trails of green criminology.

Naturally, the tool that is represented by criminal law is not enough to deal adequately with this kind of crime – to reduce the social and environmental harm deriving from it. Neither would be a mere improvement of the existing laws. Contemporary societies need a transformative process, both individual and collective, in order to succeed in 'going beyond *paranoia*, beyond that fixation that does not allow finding an escape' (Resta, 2004: 123), with the aim of promoting possible reflections and re-elaborations that are directed towards change (*metànoia*) (Resta, 2008: 171; Natali, 2014). A profound and progressive social transformation seems inevitable. The idea is to gradually build a more plausible version of the 'global theatre' in which we live, characterized by the predicament that 'facts' and 'opinions' meld with each other (Latour, 2011: 7). Rather than attempting to isolate the world of science from society and politics – as in an apotropaic ritual of purification – it is necessary to 'assemble' a political body that is capable of affirming its own responsibility in the process of change of the Earth (Latour, 2011: 7–8).

The importance of 'critical', 'radical' or 'Marxist' criminological currents resides in their interest and propensity for analysing the social reality of crime and deviance without being inevitably and automatically on the side of institutionalized power – vital if one means to contribute to social change – and this proclivity appears often in the work of green criminologists (see also Sollund, 2012). In this respect, it is important to remember that 'tradition' means, in its etymological sense, to deliver something to someone: it is never static, but dynamic, and it fosters the expansion of imagination through continuous recoveries, recognitions and renovation of the original sense (Toscano, 2008: 18). Some of the theoretical instruments that are developed by critical criminologists are in this way borrowed and applied – after being opportunely re-adapted – to the relationship between human beings and the natural environment. As Bernard Harcourt (2006: X) points out, the actual choice of adopting a determinate theoretical and methodological approach does not rest on a 'neutral' scientific decision; it is rather an ethical option with consequences and costs (in social and ecological terms) that are relevant to both society and the individual. As criminologists, it would make little sense to study environmental harm if we did not consider it to be injurious to aspects of the socioecological worlds that are really valuable (Ward, 2013: 74; Ward, 2009: 34; Becker, 2014). Putting the critical root of criminology in constant dialogue with the open and multiple perspective of green criminology could help the 'criminological mind' to provide a new twist to our understanding (Felices-Luna, 2013: 193) in facing the complex phenomenon of environmental destruction.

Acknowledgements

I would like to thank Avi Brisman and Ragnhild Sollund, who made a number of insightful and valuable suggestions in their comments on the original version of this chapter. Mirella Giulidori helped me in the translation of the text.

Notes

1. Starting from an empirical investigation carried out in the town of Huelva (Spain) on the social perception of environmental pollution (Natali, 2010, 2014), I intended to enhance the importance of the victims' perspective in the study of those corporate crimes that produce social and environmental harms. In brief, the study highlighted the multiplicity of perspectives and interpretations that the environmental victims developed about their experiences of pollution. The study also threw light on the ambiguity that links the victims' narrations on the theme of pollution to the vocabulary of justifications employed by the perpetrators of environmental crimes (see also Altopiedi, 2011).

2. Gramsci's notion of folklore, understood as a phenomenon to be studied and taken seriously (Angioni, 2011: 206–213), could also be useful in the environmental field. Today, in fact, we have to deal with new forms of subordination, with new aspects of 'common sense' that directly concern man–environment relationships. The idea of 'folk green criminology' is consonant with the central notion of 'indigenous knowledge' (see White, 2011: 117–121).

3. See also Ulrich Beck (2007). Here I refer to the notions of co-production of knowledge and post-normal science. See Tallacchini (2012).
4. See also the concept of 'solastalgia', which was conceived by Australian philosopher Albrecht (2005).
5. See the notion of 'inverted quarantine', which is promulgated by Szasz (2007) and further developed by Brisman and South (2013, 2014).

References

Albrecht, G. (2005) 'Solastalgia, a new concept in human health and identity', *Philosophy Activism Nature*, 3: 41–55.
Altopiedi, R. (2011) *Un caso di criminalità d'impresa: l'Eternit di Casale Monferrato*. Torino: L'Harmattan Italia.
Angioni, G. (2011) *Fare, dire, sentire. L'identico e il diverso nelle culture*. Nuoro: Il Maestrale.
Athens, L. (1994) 'The self as a soliloquy', *The Sociological Quarterly*, 35(3): 521–532.
Auyero, J. and Swistun, D.A. (2009) *Flammable. Environmental Suffering in an Argentine Shantytown*. Oxford, New York: Oxford University Press.
Baratta, A. (2006) '*Nomos* e *Tecne*. Materiali per una cultura post-moderna del diritto. Presentazione di Michele Marchesiello', *Studi sulla questione criminale*, 1(2): 59–65.
Beck, U. (2007) *World at Risk*. Cambridge: Polity Press.
Becker, H. (1982) *Art Worlds*. Berkeley, LA, London: University of California Press.
Becker, H. (2014) 'What about Mozart? What about murder?'. Accessed on May 2014 at http://home.earthlink.net/~hsbecker/articles/mozart.html.
Bisschop, L. and Vande Walle, G. (2013) 'Environmental victimisation and conflict resolution: A case study of e-waste'. In Walters, R., Westerhuis, D. and Wyatt, T. (eds) *Emerging Issues in Green Criminology. Exploring Power, Justice and Harm*. Hampshire, UK and New York: Palgrave Macmillan.
Bloch, E. (1959/1986) *The Principle of Hope*. Oxford: Blackwell.
Brisman, A. (2012) 'The cultural silence of climate change contrarianism'. In White, R. (ed.) *Climate Change from a Criminological Perspective*. New York: Springer.
Brisman, A. (2013) 'The violence of silence: Some reflections on access to information, public participation in decision-making, and access to justice in matters concerning the environment', *Crime, Law and Social Change*, 59(3): 291–303.
Brisman, A. and South, N. (2013) 'A green-cultural criminology: An exploratory outline', *Crime Media Culture*, 9(2): 115–135.
Brisman, A. and South, N. (2014) *Green Cultural Criminology. Constructions of Environmental Harm, Consumerism and Resistance to Ecocide*. London and New York: Routledge.
Brown, S. (2006) 'The criminology of hybrids. Rethinking crime and law in technosocial networks', *Theoretical Criminology*, 10(2): 223–244.
Butler, J. (2013) *A chi spetta una buona vita?*. Roma: Nottetempo.
Castells, M. (2001) *The Internet Galaxy: Reflections on the Internet, Business, and Society*. Oxford: Oxford University Press.

Ceretti, A. and Cornelli, R. (2013) *Oltre la paura. Cinque riflessioni su criminalità, società e politica*. Milano: Feltrinelli.

Cianchi, J.P. (2013) *I Talked to My Tree and My Tree Talked Back: Radical Environmentalists and Their Relationships with Nature*. PhD thesis, University of Tasmania. Accessed on 14 August 2014 at http://eprints.utas.edu.au/17465/.

Cohen, S. (2001) *States of Denial: Knowing About Atrocities and Suffering*. Cambridge: Polity Press.

Cottino, A. (2005) *'Disonesto ma non criminale'. La giustizia e i privilegi dei potenti*. Roma: Carocci.

Cullinan, C. (2011) *Wild Law: A Manifesto for Earth Justice*. Vermont: Chelsea Green Publishing.

Davies, P., Francis, P. and Wyatt, T. (eds) (2014) *Invisible Crimes and Social Harms*. Basingstoke: Palgrave Macmillan.

Ellefsen, R., Sollund, R. and Larsen, G. (eds) (2012) *Eco-global Crimes: Contemporary Problems and Future Challenges*. England: Ashgate Publishing.

Felices-Luna, M. (2013) 'El retorno de lo político: la contribución de Carl Schmitt a las criminologías críticas', *Revista Crítica Penal y Poder*, 5: 186–205.

Fiandaca, G. (2013) *Sul bene giuridico. Un consuntivo critico*. Torino: Giappichelli.

Greife, M.B. and Stretesky, P.B. (2013) 'Crude laws. Treadmill of production and state variations in civil and criminal liability for oil discharges in navigable waters'. In South, N. and Brisman, A. (eds) *Routledge International Handbook of Green Criminology*. London and New York: Routledge.

Hall, M. (2013) 'Victims of environmental harms and their role in national and international justice'. In Walters, R., Westerhuis, D. and Wyatt, T. (eds) *Emerging Issues in Green Criminology. Exploring Power, Justice and Harm*. Basingstoke: Palgrave Macmillan.

Halsey, M. (2013) 'Conservation criminology and the "General Accident" of climate change'. In South, N. and Brisman, A. (eds) *Routledge International Handbook of Green Criminology*. London and New York: Routledge.

Harcourt, B. (2006) *Language of the Gun. Youth, Crime and Public Policy*. Chicago and London: The University of Chicago Press.

Hillyard, P., Pantazis, C., Tombs, S. and Gordon, D. (eds) (2004) *Beyond Criminology: Taking Harm Seriously*. London: Pluto Press.

Kane, S.C. (2004) 'The unconventional methods of cultural criminology', *Theoretical Criminology*, 8(3): 303–321.

Kane, S.C. and Brisman, A. (2013) 'Technological drift and green machines: A cultural analysis of the *prius paradox*', *CRIMSOC: The Journal of Social Criminology. Special Issue: 'Green Criminology'*, Autumn 2013: 101–129.

Langer A. (1961–1995) *Il viaggiatore leggero. Scritti 1961–1995*. Palermo: Sellerio.

Larsen, G. (2012) 'The most serious crime: Eco-genocide concepts and perspectives in eco-global criminology'. In Ellefsen, R., Sollund, R. and Larsen, G. (eds) *Eco-Global Crimes: Contemporary Problems and Future Challenges*. England: Ashgate Publishing.

Lasslett, K. (2013) 'Más allá del fetichismo del Estado: el desarrollo de un programa teórico para los estudios sobre crímenes de Estado', *Revista Crítica Penal y Poder*, 5: 90–114.

Latour, B. (2011) 'Waiting for Gaia. Composing the common world through arts and politics'. Accessed on 14 August 2014 at http://www.bruno-latour.fr/sites/default/files/124-GAIA-LONDON-SPEAP_0.pdf.

Lynch, M. and Stretesky, P. (2003) 'The meaning of green: Contrasting criminological perspectives', *Theoretical Criminology*, 7(2): 213–238.

Lynch, M. (2013) 'Reflections on green criminology and its boundaries. comparing environmental and criminal victimization and considering crime from an eco-city perspective'. In South, N. and Brisman, A. (eds) *Routledge International Handbook of Green Criminology*. London and New York: Routledge.

Lynch, M., Long, M., Barrett, K. and Stretesky, P. (2013) 'Why green criminology and political economy matter in the analysis of global ecological harms', *British Journal of Criminology*, 53: 997–1016.

Mawby, R. and Walklate, S. (2000) *Critical Victimology: International Perspectives*. London: Sage.

Mead, G.H. (1963 [1934]) *Mind, Self and Society: From the Standpoint of a Social Behaviorist*. Chicago: The University of Chicago press.

Mol, H. (2013) ' "A Gift from the Tropics to the World": Power, harm, and palm oil'. In Walters, R., Westeerhuis, D.S. and Wyatt, T. (eds.) *Emerging Issues in Green Criminology. Exploring Power, Justice and Harm*. Basingstoke: Palgrave Macmillan.

Natali, L. (2010) 'The big grey elephants in the backyard of Huelva, Spain'. In White, R. (ed.) *Global Environmental Harm. Criminological Perspectives*. Devon: Willan Publishing.

Natali, L. (2013a) 'The contemporary horizon of green criminology'. In South, N. and Brisman, A. (eds) *Routledge International Handbook of Green Criminology*. London and New York: Routledge.

Natali, L. (2013b) 'Exploring environmental activism. A visual qualitative approach from an eco-global and green-cultural criminological perspective', *CRIMSOC: The Journal of Social Criminology. Special Issue: 'Green Criminology'*, Autumn 2013, Waterside Press, 64–100.

Natali, L. (2014) 'Green criminology, victimización medioambiental y social harm. El caso de Huelva (España)', *Revista Crítica Penal y Poder*, 7: 5–34.

Natali, L. (2015) *Green Criminology. Prospettive Emergenti Sui Crimini Ambientali*. Torino: Giappichelli.

Passas, N. (2005) 'Lawful but awful: "Legal Corporate Crimes" ', *The Journal of Socio-Economics*, 34(6): 771–786.

Pulcini, E. (2013) *Care of the World: Fear, Responsibility and Justice in the Global Age*. New York and London: Springer.

Resta, E. (2000) 'Contrattempi'. In Vittoria, V. (ed.) *Il futuro è già dato?*. Napoli: Alfredo Guida Editore.

Resta, E. (2004) *Il diritto fraterno*. Roma-Bari: Laterza.

Resta, E. (2008) *Diritto vivente*. Roma-Bari: Laterza.

Settis, S. (2010) *Paesaggio Costituzione Cemento. La battaglia per l'ambiente contro il degrado civile*. Torino: Einaudi.

Sini, C. (2009) *L'uomo, la macchina, l'automa*. Torino: Bollati Boringhieri.

Sollund, R. (2012) 'Introduction'. In Ellefsen, R. Sollund, R. and Larsen, G. (eds) *Eco-global Crimes: Contemporary Problems and Future Challenges*. UK: Ashgate Publishing.

South, N. (1998) 'A green field for criminology? A proposal for a perspective', *Theoretical Criminology*, 2(2): 211–234.

Stretesky, P.B., Long, M.A. and Lynch, M.J. (2013) *The Treadmill of Crime. Political Economy and Green Criminology*. Abingdon: Routledge.

Szasz, A. (1994) *Ecopopulism. Toxic Wasfe and the Movement for Environmental Justice*. Minneapolis: University of Minnesota Press.

Szasz, A. (2007) *Shopping Our Way to Safety. How We Changed from Protecting the Environment to Protecting Ourselves*. Minneapolis: University of Minnesota Press.

Tallacchini, M. (2012) 'Scienza e diritto. Prospettive di co-produzione', *Sociologia del diritto*, XXXII(1): 75–106.

Tombs, S. (2013) 'Trabajando para el mercado "libre": Complicidad estatal en la rutina del daño corporativo en el Reino Unido', *Revista Crítica Penal y Poder*, 5: 266–290.

Toscano, M.A. (2008) 'Beni culturali e sociologia'. In Toscano, M.A. and Gremigni, E. (eds) *Introduzione alla sociologia dei Beni Culturali. Testi antologici*. Firenze: Le Lettere.

Vasta, N. (2010) 'Negotiating roles and identities in corporate advertising: A multimodal analysis of the total energy doubled 2005 TV commercial'. In Swain, E.A. (ed.) *SFL at the Frontier: Thresholds and Potentialities of SFL as a Descriptive Theory*. Trieste: EUT.

Walters, R. (2010) 'Toxic atmospheres air pollution, trade and the politics of regulation', *Critical Criminology*, 18: 307–323.

Walters, R., Westerhuis, D.S. and Wyatt, T. (2013) 'Introduction'. In Walters, R., Westeerhuis, D.S. and Wyatt, T. (eds) *Emerging Issues in Green Criminology. Exploring Power, Justice and Harm*. Basingstoke: Palgrave Macmillan.

Ward, T. (2009) 'Antiquities, forests, and Simmel's sociology of value'. In Mackenzie, S. and Green, P. (eds) *Criminology and Archaeology. Studies in Looted Antiquities*. US and Canada: Hart Publishing.

Ward, T. (2013) 'El crimen de Estado y la sociología de los Derechos Humanos', *Revista Crítica Penal y Poder*, 5: 63–76.

White, R. (2003) 'Environmental issues and the criminological imagination', *Theoretical Criminology*, 7: 483–506.

White, R. (2008) *Crimes against Nature: Environmental Criminology and Ecological Justice*. London: Willan Publishing.

White, R. (2011) *Transnational Environmental Crime: Toward an Eco-global Criminology*. London and New York: Routledge.

White, R. (2013) 'Resource extraction leaves something behind: Environmental justice and mining', *International Journal for Crime and Justice*, 2(1): 50–64.

White, R. and Heckenberg, D. (2014) *Green Criminology. An Introduction to the Study of Environmental Harm*. London and New York: Routledge.

Williams, C. (1996) 'An environmental victimology', *Social Justice*, 23: 16–40.

Williams, C. (1998) (ed.) *Environmental Victims. New Risks, New Injustice*. London: Earthscan.

Zaffaroni, E.R. (2012) *La Pachamama y el humano*. Buenos Aires: Ediciones Madres de Plaza de Mayo.

5

'Creative Destruction' and the Economy of Waste

Vincenzo Ruggiero

Introduction

The environment has been among the theoretical and practical concerns of criminology for many years, but it could be argued that such concerns have long been 'indirect' in nature.[1] The real criminological object of study in past decades was how organized and white-collar individuals operated in illicit businesses that had an environmental impact – for instance, companies related to garbage disposal or the construction industry. The former activity, as was often found, was performed outside the statutory rules that established the types of substance to be dumped and exactly where they were to be disposed of. The latter, as investigators proved, led to illicit building on geologically hazardous sites, and contrary to the guidelines regarding the precise materials to be used and their proven safety. Not surprisingly, investigators and scholars who focused on these issues were experts in organized and white-collar crime, with the environment providing a mere backdrop for law enforcement and academic efforts (Arlacchi, 1983; Abadinsky, 1990; Block, 1991; Gambetta, 1992; Ruggiero, 1996). It was only in the 1990s that a proper 'green field' for criminology began to develop (Lynch, 1990; Berat, 1993; Gray, 1996; South, 1998), and this chapter attempts to add to the analysis and information that has been produced over the last two decades in that field. The first section focuses on the illegal dumping of waste. The description of this specific form of eco-crime is followed by a brief outline of the main concerns around environmental harm that is caused not only by illegitimate but also by legitimate behaviour. The subsequent analysis, after defining the legal framework in which environmental harm is caused, draws on the categories that were utilized by Max Weber in his study of the relationship between law and economy. A discussion of the variables 'innovation' and 'deviance', which are used in economics as well as in criminology, concludes the article, which finally examines the very logic of economic development. This logic may explain the prevalence of environmental crime, while alternative economic analysis, it is argued, may provide the tools for its prevention.

Dirty-collar crime

Research that has been conducted over recent decades has shown that processing industrial waste without a licence and sidestepping environmental regulations 'is cheaper and faster'. Cases that have been uncovered in several countries prove that illegal enterprises may offer service packages which comprise false invoices, transport facilities, mendacious chemical reports as to the nature of the substances dumped and forged permits to dump. The dynamics of this illicit activity are similar in most parts of the world and were already well documented two decades ago (van Duyne, 1993; Brants, 1994; Moore, 1994). More recently it has been found that, in European countries, some legally registered companies also operate illegally. They either establish partnerships with legitimate firms or run their own in-house, parallel, illicit business (Ruggiero, 2010; Ruggiero and South, 2010). The choice between the two services is determined by how much the customer is prepared to pay. It is otiose, in this respect, to question whether customers are aware of the illegal nature of the cheaper option because its very cheapness speaks for itself (Mandel, 1999; Liddick, 2010). Mandel (1999: 66) describes the business of such violators as 'unsanctioned hazardous materials transfers', moving unwanted, frequently toxic, waste from regulated spaces to sites where weak or no opposition will be encountered and from developed to developing nations, all part of a global industry of various 'deadly transfers' that occur across a 'disorderly world'. In the USA, research indicates that the involvement of organized crime reaches all aspects of the business: the control of which companies are officially licensed to dispose of waste, which earn contracts with public or private organizations, the payment of bribes to dump-site owners, and the management of such sites (Block and Scarpitti, 1985; Szasz, 1986; Salzano, 1994; Liddick, 2010).

Recent examples in Germany show that even in countries where the legislation is progressive and clear, the illegal disposal of waste is widespread. Such cases emerged when a mismatch was noted between the quantity of waste expected and what was actually received by incinerators operating in the eastern regions of the country. The missing portion of waste was found to have been dumped at illegal disposal sites (Natale, 2009). Entrepreneurs who were using such dumps opted for the cheapest means of waste management, thus circumventing the rules, which impose a fee of around €200 per tonne of waste treated. Cases also emerged in which the composition of the waste treated was falsely certified, so that substances which should have been disposed of at special sites were instead dumped at inappropriate ones. That cases such as these occur in highly ecologically aware Germany may be surprising. However, the paradox is that the development of illegal dumping services runs parallel with the very increase in environmental awareness, the latter forcing governments to raise costs for industrial dumping, which indirectly encourages industrialists to opt for cheaper, if illicit, solutions. Moreover, the logic of illegal dumping resembles that of arms production: the accumulation of weapons in times of peace is meant to provide an immediate supply when they are suddenly deemed necessary. However, their very availability makes the resource to war more likely or even a constant possibility. Constructing illegal dumps, similarly, may not be the result of specific demand by entrepreneurs but may contribute to triggering that demand once illegal dumping facilities are ready available. Finally, dumping waste abroad, in developing countries, is among the cheapest solutions and can be described as a form of ecological racism.

Past and current cases of illegal waste disposal share a key characteristic – namely, they are the result of partnerships between the official economy and illicit enterprise. When we think about these partnerships, we tend to identify one of the actors involved in the transaction with organized crime. This is true in many cases, in the sense that organized crime may offer a service to legitimate business (Ruggiero, 1996, 2000; Gounev and Ruggiero, 2012). However, it may be at times a redundant actor in this business, or may be encouraged to intervene due to the ineffectiveness of both states and entrepreneurs. The notorious case of Naples, in this respect, is an example worth reassessing.

The 'rubbish crisis' that has been affecting Naples over recent years involved, first of all, industrial managers and public administrators: the former defrauded the public administration, while the latter proved incapable of controlling the work of those that they commissioned and failed to denounce the fraud. Judges brought charges against the company Impregilo, which was entrusted with the construction of a multilayered disposal and recycling system but failed to do so. This failure led to the well-known emergency situation, and with a waste of money that is quantifiable at about €8 billion: 60 tons of rubbish were scattered on the streets of Naples province. Managers at Impregilo were accused of presenting an inadequate and fraudulent tender while being aware that the price quoted was unrealistic and that their company lacked the technical capacity to perform the job that was required. The mayor of Naples, on the other hand, was charged with gross negligence and complicity in the fraud, having granted an invalid contract and failed to intervene when the improper conduct of the beneficiary became manifest (Piccoli, 2008). The judicial investigation was a clear response to widespread stereotypes, particularly that responsibility for the rubbish crisis was to be directly attributed to organized crime. Organized crime, in fact, only intervened when the chaotic situation that was caused by entrepreneurs became manifest. But it would be wrong to impute all of the illegality displayed in this case to structured, traditional organized criminal groups. Improvised businessmen started to buy land from small farmers to turn it into illegal dumps, while improvised lorry owners limited their role to the transportation of garbage. The complicity of local politicians was detectable in the hasty, routine authorizations that were given to such unlikely entrepreneurs. Assumptions about one single, stifling, violent organization encompassing myriads of illegal acts under a nightmarish 'Gomorrah' totally miss the mark (De Crescenzo, 2008; Ruggiero and South, 2010; Ruggiero, 2010, 2013).

Hazardous productions can be endorsed by governments, which underplay the dangers to people and the environment because, at least officially, they intend to protect key sectors of the economy and important sources of employment opportunities. In some cases the economy and employment are not the main concerns of state agents, who turn a blind eye in the face of dangerous productions in exchange for bribes. In association with lobbyists, these state agents establish alliances and partnerships which mimic those that commonly characterize the activities of organized criminals. We are now entering the arena of environmental harm caused by 'legitimate' behaviour.

Harm as crime

Many novel issues and conducts have compounded the dilemmas and expanded the arena of green criminology in ways that not long ago would have been unpredictable. Climate change, the disposal of toxic waste, illegal fishing, deforestation, the exploitation of tar sands, the illegal trade in reptiles and endangered species and the destruction of biodiversity are only some of them. What these acts have in common is an international, global character. What differentiates them is whether or not they constitute violations of law. The notion of environmental harm which denotes this field of research transcends the legal definitions that are provided by the jurisprudence and by conventional criminology, and it relates specifically to the wider ecological and green domain. The task of what is termed 'eco-global criminology' is therefore to name harms as criminal, irrespective of their legal definition (White, 2010). In this way, the traditional perspective that guides the study of white-collar crime returns, as found in the pioneering work of Edwin Sutherland.

Research has addressed equatorial deforestation as a harmful practice and a criminological issue. In a 'world tour' around the equator, three types of, often illegal, activity that characterize the deforestation of tropical rainforests are found: logging, mining and land conversion for agriculture. The effects of such practices may be disastrous, as they are said to be responsible for 20% of global greenhouse emissions. A typical area in which Sutherland's intuitions could easily be applied, this is an area in which criminologists may well be active while law enforcement is mostly absent (Boekhout van Solinge, 2010).

That these concerns are global is proved by the movement of toxic harms across increasingly porous borders (Heckenberg, 2010). Green criminology locates 'transfer' within the growing interconnection of markets and the expanding flow of goods, a process that is not exempt from its own specific form of 'othering' (South, 2010). In other words, the production and delivery of specific goods, services and technologies are located in particular countries, where human costs and environmental harms are also transferred. Such countries, of course, may be in the condition where priority for survival and subsistence, wittingly or otherwise, lead them to accept ecological risks. Relocation of activities or substances that cause environmental degradation to vulnerable countries and communities is deemed to be a form of ecological imperialism that relies on established 'toxic distribution networks'. A key question posed by this type of research is why the violations that are involved continue to be associated with, or officially simply defined as, harms when in lay terms they are criminal in both nature and impact. The examination of environmental harm as non-criminalized conduct also focuses on global warming as a global crime, with a particular emphasis upon the complicity between national states and corporate actors. The 'ecological footprint' of nations is discussed – namely, the amount of land, water and air used by them to produce the commodities each of them consumes. 'Despite its wealth of natural resources, the nation with the largest ecological footprint is the US, making it the largest contributor to the problem of global warming' (Lynch and Stretesky, 2010: 64).

The literature offers examples of polluting behaviour in fast-growing economies and developing countries. The case of China is, in this respect, of particular interest for the geopolitical backdrop against which it should be read: are new, rampant economies to be allowed to follow in the footsteps of developed economies, thus claiming their equal right to destroy and degrade the environment, or are they to be restrained in their development? If not, how can we still obsessively think of the variables 'development' and 'growth' as a panacea for the wellbeing of humanity? In a concluding remark to his research, Yang Shuqin (2010: 158) candidly argues:

> As a Chinese citizen, I have deeply felt the benefits brought by the recent economic development as I have grown up. However, it is also painful to see that the blue sky, green trees and clear water are leaving us. I sincerely wish that while investing, building factories and making profits in China, the multinational companies could leave us a blue sky, and our offspring a foundation for sustainable development.

How social justice can accommodate responses to environmental crime is still hard to see. Responses may derive from the drawing together of political and practical action to shape public policy, beyond the state-territorial principle (South, 2010). On the other hand, the ecocidal tendencies of late modernity may be hard to oppose, as the green movement has repeatedly experienced. Green criminology, it would appear, is destined to encounter the same dilemma that has hampered green party politics for many years. The former may well be concerned with harm as the outcome of both legal and illegal practices, and invoke notions of environmental morality and ecological rights. However, it is bound to relate its research and intellectual production to the vexed distinction between 'shallow' and 'deep' ecology. Shallow ecology appears to believe that the technology which is destroying the environment may also rescue it: a managerial approach to environmental problems will be sufficient to solve problems, without fundamental changes in present values or patterns of production and consumption. Deep ecology, by contrast, embraces a holistic outlook, whereby humans are interconnected with each other and are constantly in relationship with everything around them – they are part of the flow of energy, the web of life (Bookchin, 1980, 1982). Radical changes in production and consumption patterns, but also in the fundamental principles and values that are expressed by the undeservedly respected 'science' of economics, are necessary.

It is worth, now, discussing the legal framework within which activities that harm the environment take place.

A conceptual hybrid

Environmental law can be described as a conceptual hybrid in that its doctrinal content largely derives from principles that are enunciated in other legal contexts (Ruggiero and South, 2010). It is inspired, on the one hand, by public law, which consists of sets of regulations, procedural constraints and control processes. It contains, on the other hand, elements of private law, where it affects property and other recognized rights and interests. 'Therefore, there can be a sense that environmental law discourse is ultimately shackled by a dependent, satellite status, a repository of greener values, but for the most part swimming against a distinctively ungreen tide of prevailing legal priorities' (Stallworthy, 2008: 4–5). Environmental law, in other words, suffers the legacy of legal reasoning that is geared to the protection of socioeconomic systems that are heavily orientated towards unfettered industrial growth, production and consumption.

Increasing commitment to market freedom has created a situation in which ethics, education and the 'invisible' mechanisms of the economy itself are seen as the only regulatory tools upon which states are expected to rely. However, critics argue that legal control cannot be discarded, and that strategies require 'legal embeddedness' if they are to succeed. 'The environment needs good law if it is to avoid suffering further serious harm' (Wilkinson, 2002: 8). Specifically, laws on waste disposal are faced with the challenges that are posed by the following three categories of conduct: (1) legal persons discharging substances in accordance with the conditions that are established by a licence; (2) legal persons discharging substances in breach of their licence; and (3) legal persons discharging substances without holding a licence (Wolf and Stanley, 2003). It may be true, as Stallworthy (2008: 1) argues, that environmental law is evolving 'to the stage that it has developed a coherent basis of applicable theory and principles'. It has to be stressed, however, that such law has mainly focused upon the second and third category mentioned above – namely, the harm caused by white-collar, corporate or conventional offenders – while the damage caused by industrial development has remained largely unaddressed. And yet the reach of environmental law could potentially introduce into legal discourse 'long unasked questions as to the ecosystem and biodiversity protection, as well as appropriate conditions for access and use of natural resources' (ibid: 3).

In response to such problems, the notion of intergenerational equity has been set forth – namely, a theory of 'justice between generations' that identifies obligations and rights that are enforceable in international law (White, 2010). According to this theory, each generation receives a natural and cultural legacy from previous generations that it holds in trust for succeeding ones. This partnership between the living, the dead and the unborn entails 'a duty on mankind to pass on to succeeding generations a planet at least as healthy as the one it inherited so that each generation will be able to enjoy its fruits' (Kofele-Kale, 2006: 324). However, it is hard to establish how such a moral obligation ought to be turned into a legal one. Some authors tend to see its fairness and concerns as perfectly suitable for incorporation into statutory legal principles (Wolf and Stanley, 2003). Others, by contrast, criticize governments for their unwillingness or inability to translate such moral obligation into radical regulatory measures. In an effort to balance business interest with public interest, governments can at most implement policies that limit rather than eliminate environmental damage. Such measures may include the 'polluter pays' rule, whereby businesses should internalize the costs of the pollution that they generate; 'eco-taxes', which are expected to encourage firms to reduce the environmental impact of their activities; and 'emissions trading' as an 'eco-instrument' which establishes the maximum level of 'pollution credits' for businesses. 'Over time, the regulator reduces the number of credits in circulation and this results in an increase in the price of the credits. This provides a financial incentive for participating firms to reduce their need for credits by developing less polluting methods of production' (ibid: 18).

Critics of these eco-instruments remark that environmental law as a whole has proved to be a colossal failure, despite good intentions and the hard work of many citizens, lawyers and government officials. Agencies are accused of adopting an excessive degree of discretion in their statutes so that continuing damage to the atmosphere and other natural resources is allowed. In response, a 'public trust doctrine' is advocated as a fundamental mechanism to ensure governmental protection of the environment and of public welfare. 'At the core of this doctrine is the principle that every sovereign government holds vital natural resources in "trust" for the public'. In this way a shift is encouraged from a system that is driven by political discretion to 'one that is infused with public trust principles and policies across all branches of government and at all jurisdictional levels' (Wood, 2009: 43). The expansion of the public's *res* would add new quantifiable assets to the range of collective protected interests. 'While the courts have traditionally focused on water and wildlife resources in applying the public trust, the new climate-altered world demands a far more encompassing definition of the public's natural *res*' (ibid: 78).

A more critical approach to this topic emerges when Max Weber's analysis of law and the economy is revisited. The broader perspective provided by him, as I will attempt to show, leads to a deeper understanding of the relationship between legality, economic development and the environment.

Law and economy

There are legal and sociological points of view. Through the former we ask: What is intrinsically valid as law? That is to say: What significance or what normative meaning ought to be attributed to a verbal pattern that has the form of a legal proposition? From a sociological perspective, the question becomes:

> What *actually* happens in a group owing to the *probability* that persons engaged in social action, especially those exerting a socially relevant amount of power, subjectively consider certain norms as valid and practically act according to them, in other words, orient their own conduct towards these norms?
>
> (Weber, 1978: 311)

This distinction also determines the relationship between law and economy.

Jurists, Weber explains, take for granted the empirical validity of legal propositions, therefore they examine each of them and they try to determine the logic and the meaning that those propositions have within a coherent system. Jurists, in brief, are concerned with the 'legal order'. Sociological economics, on the other hand, considers actual human activities as they are conditioned by the necessity to take into account the facts of economic life. 'The legal order of legal theory has nothing directly to do with the world of real economic conduct, since both exist on different levels. One exists in the realm of the *ought*, while the other deals with the world of the *is*' (ibid). Environmental crime and the illegal dumping of waste belong to the world of the *is*, and they constitute strategies that are prompted by the difficulties and uncertainties that actors face in the economic arena. Such strategies spread, multiply and mushroom, becoming what Weber describes as 'habituation' and 'custom', and, although illicit, they determine 'unreflective conducts' which, with time, become morally acceptable.

Mere custom can be of far-reaching economic significance in that its strategies, as Weber suggests, do not arouse the slightest disapproval while gradually giving way to imitation. 'Adherence to what has as such become customary is such a strong component of all conduct and, consequently, of all social action, that legal coercion, where it opposes custom, frequently fails in the attempt to influence actual conduct' (ibid: 320). Convention is equally effective, if not more so. Weber argues that individuals are affected by responses to their action that emanate from their peers rather than from an earthly or transcendental authority. The existence of a 'convention' may thus be far more determinative of conducts than the existence of a legal enforcement machinery. Of course, it is hard to clearly establish the point at which certain actions become custom and certain modes of conduct take on a binding nature. Nevertheless, Weber sees 'abnormality' (or, in our lexicon, deviance) as the 'the most important source of innovation ... capable of exercising a special influence on others' (ibid: 320–321).

Innovation, in economic initiative as well as in deviance, is constituted by a number of elements. First there is inspiration – namely, a sudden awareness that a certain action ought to be undertaken, irrespective of the drastic or illegitimate means that it requires. Second there are empathy or identification – namely, influencing others into acting and at the same time being influenced by their action. When conducts begin to observe a degree of regularity, the third element – oughtness – emerges, 'producing consensus and ultimately law' (ibid: 323).

> Obviously, legal guarantees are directly at the service of economic interests to a very large degree. Even where this does not seem to be, or actually is not, the case, economic interests are among the strongest factors influencing the creation of law ... The power of law over economic conduct has in many respects grown weaker rather than stronger.
>
> (ibid: 334–335)

Weber is suggesting that the relationship patterns among economic actors are determined by experimentation that then turns into habit and is impervious to normative adjustments. The difficulties increase with the degree of development and the growing interdependence of individual economic units in the market and, consequently, the dependence of everyone upon the conduct of others. But, crucially, Weber remarks that the limitation of successful legal coercion in the economic sphere is due to the strength of private economic interests on the one hand, and interests promoting conformance to the rules of law on the other. The inclination to forego economic opportunity simply in order to act legally is obviously slight, he remarks, unless circumvention of the formal law is strongly disapproved of by powerful actors and collectivities. 'Besides, it is often not difficult to disguise the circumvention of a law in the economic sphere' (ibid: 335).

In brief, looking at environmental crime, we are faced with economic subjects that are engaged in instrumentally rational action, which is determined by the calculated end profit; value-rational action, which is determined by the belief that bending rules is an ethical necessity; affectual action, which is inspired by the emotional aspect of material gain; and traditional action, which is 'determined by ingrained habituation' (ibid: 24–25).

Weber describes a process whereby deviance is a constant possibility in economic initiative, as his analysis embraces both legitimate and illegitimate profit-making activities. His focus on opportunities for predatory profit, for example, explicates the relationship between economic actors and politicians, remarking that the latter are not required to design legal rules in general but legal rules which maximize economic efficiency and profit. In his analysis, therefore, innovation amounts to the violation of norms, while habit establishes the regularity of new conducts. In this process it is law which is required to adapt to innovation rather than the other way round. Let us see some aspects of this process in more detail.

Innovation as deviance

In economic classical thought we often find warnings about an in-built tendency that determines a decline in profits. Against this tendency, innovation is advocated whereby entrepreneurs are expected to mobilize their creativity and, in perpetual agitation, transcend established conducts, in a process sustained by constant transgression. For David Ricardo (1992), for example, innovation sums up the difficulties, ambiguities and shortcomings of economic initiative and its inherent transgressive impetus. The concept is fully developed by Schumpeter (1961a, 1961b), who identifies the main characteristics of the entrepreneurial spirit exactly on the basis of the variable innovation. He distinguishes between those economic actors that passively follow tradition and those that are more inclined to adopt new technologies. Only the latter are granted the definition of entrepreneurs, as the economic process, in his view, is an evolutionary one, and when forced to remain stationary it should not be described as a process in the first place. The fundamental impulse that sets and keeps the economic engine in motion, according to Schumpeter, derives from new consumer goods, new methods of production or distribution, new markets and new forms of industrial organization. He resorts to a biological metaphor to illustrate the process of economic mutation. This, he argues, incessantly revolutionizes the economic structure from within, incessantly destroying the old one, incessantly creating a new one. This is the celebrated concept of 'creative destruction' that was elaborated by Schumpeter.

It is not surprising that the term 'innovation' found its way into the vocabulary of the sociology of deviance. The term, in effect, while capturing the entrepreneurial spirit in a nutshell, also encapsulates a disquieting gist of entrepreneurial deviance. Economic actors, in order to be actors at all, must avoid the habitual flow, escape from stagnant conditions and deviate from mainstream behaviour: they must fight against the whirl of conformity. These observations, made by economist Schumpeter, echo those that were elaborated within the sociology of deviance by Merton (1968). Innovation, for Merton, is one of the deviant adaptations that are available to strained social and economic conditions. In his words, the history of the great American fortunes is threaded with various strains towards institutionally dubious innovation, while of those located in the lower reaches of the social structure the culture makes incompatible demands. On the one hand they are asked to pursue wealth and success, and on the other they are largely denied effective opportunities to do so legally. Within this context, with society placing a high premium on affluence and social ascent, and with the channels of vertical mobility being closed or narrowed, Al Capone represents the triumph of amoral intelligence over morally prescribed failure. But let us take the analysis of the variable innovation a bit further.

If this variable epitomizes the ambiguity of 'economic development' and 'crime' as discreet spheres of human activity, how does it apply to the sphere of environmental crime? These crimes innovate both in Schumpeter's sense and in Merton's. They introduce new combinations of productive factors while devising deviant adaptations to economic strain, therefore pursuing legitimate goals through illegitimate means.

Creative destruction

While for Max Weber innovation results from, and is the result of, the distance between law and economy, for Schumpeter it alludes to a process of creative destruction, as we have seen, which in his view is the essential fact about market economies. Economists would retort that the deviance that was identified by Weber in the economic process, and the destruction posited by Schumpeter in the form of innovation, amount to externalities – namely, unintended consequences that are suffered by third parties (i.e. individuals and groups that do not participate directly in a transaction). Such consequences are not among the preoccupations of economists; rather, they belong to the remit of states. In this way, states are not required to interfere with market forces but are called upon only at the final stage of the economic process – that is, when the devastation that is produced by such a process becomes visible and has to remedied.

In this final section, a crucial aspect of environmental crime may emerge if the implicit logic of economic development is critiqued. According to the picture provided by Max Weber, we are faced with conducts which are hardly susceptible to the control and discipline of legal norms. If we adhere to this view in a more comprehensive way, we have to conclude that development itself, and the growing complexity of markets, make legal coercion increasingly difficult to apply to the economic sphere. As a logical consequence, we may advocate a halt to economic development as the only way of reducing and preventing environmental crime. This is an argument against insatiability put forward by a number of critical economists who say 'enough is enough', thus challenging the current obsession with the growth of gross domestic product (GDP) (Skidelsky and Skidelsky, 2012). 'To say that my aim in life is to make more and more money is like saying that my aim in eating is to get fatter and fatter' (ibid: 5). The critique of wealth, growth and GDP may constitute a good analytical start for the designing of preventative measures.

Neoliberal thinkers such as Hayek (1973) tell us that economic initiative forges a 'spontaneous order', a utopian state of affairs to which market actors will attempt to adhere, but only rarely will they approximate. Deviant elites that harm the environment translate this utopia into concrete practice, albeit such translation requires the violation of rules and illegality. In their case, total freedom 'spontaneously' leads to crime, a form of 'creative destruction' more real than metaphorical. Such destruction targets not only institutional frameworks and traditional forms of state sovereignty but also 'social relations, welfare provisions, ways of life and thought, reproductive activities, attachment to the land and habits of the heart' (Harvey, 2011: 3). Neoliberalism, in advocating the maximization of the reach and frequency of market transactions, seeks to bring all human action into the domain of the market. The consequence of this economic theology is that markets are required to replace governments, and that economics will be entrusted with the task of abolishing politics, which is seen as a cumbersome obstacle to freedom of choice (Agamben, 2009). Economics as a 'science' posited by neoliberalism cannot accept being hindered by human and political choice, as choices are regarded as automatic, necessary outcomes of mathematical formulae, uncontrollable effects of competing actors in the marketplace (Terni, 2011).

Moreover,

> The drive towards market freedoms and the commodification of everything can all too easily run amok and produce social incoherence. The destruction of forms of social solidarity leaves a gaping hole in the social order. It then becomes peculiarly difficult to combat anomie and control the resultant anti-social behaviours such as criminality.
>
> (Harvey, 2011: 80)

Is wealth a value? This question was vehemently posed over four decades ago by Dworkin (1980), who contested the commonly shared assumption that wealth maximization is the core aim of economic initiative. He set off his argument by noting that societies with more wealth are not necessarily better off than societies with less. Only those who personify society believe that the former imply wealthier individuals. Wealth may be thought of as a component of social value – that is, something that is worth having for its own sake. But there are two versions of this claim, termed the 'immodest' and 'modest' versions, respectively. The first holds that social wealth is the only component of social value, while the second argues that it is one component among other values. The second links value with a distributional component, thereby describing wealth as an instrument that enables all 'to lead a more valuable, successful, happier, or more moral life' (ibid: 201). A safe and clean environment is, we may add, among those values that should accompany the production of wealth.

In a similar critique of the concept of growth, it is noted that its very measurement ignores variables such as distribution and social justice: countries with exceptional growth rates may display exceptional levels of inequality and an average low quality of life. Turning to the official measurement of GDP, this results from the sum of all of the goods and services produced inside a country, divided by the number of people who inhabit it (Fioramonti, 2013). Again, the measurement distorts the actual success of countries and their economic systems, not only because it fails to consider the variable social equality but also because it includes some service sectors which signal bad, rather than good, performance. For example, considering the health and public sectors, expenditure in these tells us more about the victims of growth than anything else, because those in need of health and public care are the 'unintended consequences' of a growing GDP.

> America, for instance, gets worse health outcomes, in terms of longevity or virtually any other measure of health performance, but spends more money. If we were measuring performance, the lower efficiency of America's sector would count against the US, and France's health care sector output would be higher. As it is, it's just the reverse, the inefficiency helps inflate America's GDP number.
>
> (Stiglitz, 2012: 183)

GDP does not adequately capture costs to the environment, nor does it assess the sustainability of the growth that is occurring. In fact, costs to the environment are related in a positive manner to GDP because they officially reflect entrepreneurial efforts, productive activity and wealth. On the other hand, the depletion of resources should result in diminishing wealth and a declining GDP. However, 'industries like coal and oil want to keep it that way. They don't want the scarcity of natural resources or the damage to our environment to be priced, and they don't want our GDP metrics to be adjusted to reflect sustainability' (ibid: 99). Including the costs to the environment as a negative item within GDP would imply that industries should be charged for the damage that is caused. As they are not charged, they are indirectly receiving hidden subsidies, which add to other gifts such as favourable tax treatment and access to resources at below fair market prices. Oil companies that intend to intensify or multiply offshore drilling are aware that, simultaneously, they have to ensure that laws are implemented which make them unaccountable for the possible damage that is done. 'Because of the oil and coal companies that use their money to influence environmental regulation, we live in a world with more air and water pollution, in an environment that is less attractive and less healthy than would otherwise be the case' (ibid: 99). Those who oppose economic regulations argue that they are costly and that they reduce growth. According to a critical view, instead, economic development causing environmental degradation makes a negative contribution to the creation of wealth.

Challenging growth implies a critique of consumption, which was rendered by Keynes (1978) into a critique of wants. This involves a comparison between what one wants and what others have, and a realization that no level of material wealth is likely to be satisfying as far as others possess more (Skidelsky and Skidelsky, 2012). Wants come in the form of 'status spending' – namely, consumptions which make us feel superior to our fellows, or as advertisements of our own success in accumulating money.

From this perspective, growth is criminogenic not only because it can be metaphorically equated to obesity but also because it depicts greed and acquisitiveness in a positive light, making them core values of individual and collective behaviour. Simultaneously, growth as we have experienced it over the decades exacerbates the polarization of wealth, therefore increasing relative deprivation, which is one of the central variables in the analysis of crime. Ultimately, as a manifestation of insatiability, growth is a form of pathology, like the uncontrollable desire to collect things or to swallow enormous quantities of food. A radical critique of economic growth, therefore, could be a first step towards the prevention of environmental crime.

In conclusion, a full understanding of environmental crime requires an analysis of illegal behaviour that is adopted by conventional criminal organizations, but also of the illicit practices that are put in place by official economic actors and political representatives. Finally, it attracts attention to the very logic of economic development, the 'creative destruction' that is encouraged by unfettered growth (Ruggiero, 2013). No other harmful activity requires similar multidisciplinary efforts.

Note

1. Previously published as ' "Creative Destruction" and the Economy of Waste', *Crítica Penal y Poder* 4: 34–48 (ed. Observatory of the Penal System and Human Rights, University of Barcelona) reprinted with permission.

References

Abadinsky, H. (1990) *Organized Crime*. Chicago: Nelson-Hall.
Agamben, G. (2009) *Il regno e la gloria: per una genealogia teologica dell'economia e del governo*. Turin: Bollati Boringhieri.
Arlacchi, P. (1983) *La mafia imprenditrice*. Bologna: Il Mulino.
Berat, L. (1993) 'Defending the right to a healthy environment: Towards a crime of genocide in international law', *Boston University International Law Journal*, 11: 327–348.
Block, A. (1991) *Perspectives on Organizing Crime*. Dordrecht: Kluwer.

Block, A. and Scarpitti, F.R. (1985) *Poisoning for Profit: The Mafia and Toxic Waste in America*. New York: William Morrow.
Boekhout van Solinge, T. (2010) 'Equatorial deforestation as a harmful practice and a criminological issue'. In White, R. (ed.) *Global Environmental Harm*. Cullompton: Willan.
Bookchin, M. (1980) *Toward an Ecological Society*. Montreal: Black Rose Books.
Bookchin, M. (1982) *The Ecology of Freedom*. Palo Alto: Cheshire Books.
Brants, C. (1994) 'The system's rigged – or is it?' *Crime, Law and Social Change*, 21: 103–25.
De Crescenzo, D. (2008) 'Sporco, troppo sporco', *Narcomafie*, 7–8: 15–16.
Dworkin, R.M. (1980) 'Is wealth a value?' *Journal of Legal Studies*, IX(2): 191–224.
Fioramonti, L. (2013) *Gross Domestic Problem: The Politics Behind the World's Most Powerful Number*. London: Zed Books.
Gambetta, D. (1992) *La mafia siciliana. Un'industria della protezione privata*. Turin: Einaudi.
Gounev, P. and Ruggiero, V. (eds) (2012) *Corruption and Organised Crime in Europe*. London: Routledge.
Gray, M.A. (1996) 'The international crime of ecocide', *California Western International Law Journal*, 26: 215–271.
Harvey, D. (2011) *A Brief History of Neoliberalism*. Oxford: Oxford University Press.
Hayek, F.A. (1973) *Law, Legislation and Liberty*. London: Routledge & Kegan Paul.
Heckenberg, D. (2010) 'The global transference of toxic harms'. In White, R. (ed.) *Global Environmental Harm*. Cullompton: Willan.
Keynes, J.M. (1978) *Essays in Persuasion*. Cambridge: Cambridge University Press.
Kofele-Kale, N. (2006) *The International Law of Responsibility for Economic Crimes*. Aldershot: Ashgate.
Liddick, D. (2010) 'The traffic in garbage and hazardous waste: An overview', *Trends in Organized Crime*, 13: 134–148.
Lynch, M.J. (1990) 'The greening of criminology: A perspective on the 1990s', *Critical Criminology*, 2: 11–12.
Lynch, M.J. and Stretesky, P.B. (2010) 'Global warming, global crime: A green criminological perspective'. In White, R. (ed.) *Global Environmental Harm*. Cullompton: Willan.
Mandel, R. (1999) *Deadly Transfers and the Global Playground: Transnational Security Threats in a Disorderly World*. Westport, CT: Praeger.
Merton, R. (1968) *Social Theory and Social Structure*. New York: The Free Press.
Moore, R.H. (1994) 'The activities and personnel of twenty-first century organised crime', *Criminal Organizations*, 9: 3–11.
Natale, S. (2009) 'Germania, modello Napoli', *Narcomafie*, 4: 20–25.
Piccoli, G. (2008) 'Rifiuti, camorra e mala amministrazione', *Narcomafie*, 7–8: 12–21.
Ricardo, D. (1992[1817]) *The Principles of Political Economy and Taxation*. London: Everyman's Library.
Ruggiero, V. (1996) *Organised and Corporate Crime in Europe*. Aldershot: Dartmouth.
Ruggiero, V. (2000) *Crime and Markets*. Oxford: Oxford University Press.
Ruggiero, V. (2010) 'Dirty collar crime in Naples', *UN Freedom from Fear*, March 29 (published online).

Ruggiero, V. (2013) *The Crimes of the Economy: A Criminological Analysis of Economic Thought*. London: Routledge.

Ruggiero, V. and South, N. (2010) 'Green criminology and dirty collar crime', *Critical Criminology*, 18: 251–262.

Salzano, J. (1994) 'It's dirty business: Organized crime in deep sludge', *Criminal Organizations*, 8: 17–20.

Schumpeter, I. (1961a) *Capitalism, Socialism and Democracy*. London: Allen & Unwin.

Schumpeter, I. (1961b) *The Theory of Economic Development*. New York: Oxford University Press.

Shuqin, Y. (2010) 'The polluting behaviour of the multinational corporations in China'. In White, R. (ed.) *Global Environmental Harm*. Cullompton: Willan.

Skidelsky, R. and Skidelsky, E. (2012) *How Much is Enough? The Love of Money and the Case for the Good Life*. London: Allen Lane.

South, N. (1998) 'A green field for criminology?' *Theoretical Criminology*, 2: 211–234.

South, N. (2010) 'The ecocidal tendencies of late modernity: Transnational crime, social exclusion, victims and rights'. In White, R. (ed.) *Global Environmental Harm*. Cullompton: Willan.

Stallworthy, M. (2008) *Understanding Environmental Law*. London: Sweet & Maxwell.

Stiglitz, J.E. (2012) *The Price of Inequality*. London: Allen Lane.

Szasz, A. (1986) 'Corporations, organized crime, and the disposal of hazardous qaste: An examination of the making of a criminogenic regulatory structure', *Criminology*, 24: 1–27.

Terni, M. (2011) *La mano invisibile della politica. Pace e guerra tra stato e mercato*. Milan: Garzanti.

van Duyne, P. (1993) 'Organised crime and business crime-enterprises in the Netherlands', *Crime, Law and Social Change*, 19: 103–142.

Weber, M. (1978) *Economy and Society: An Outline of Interpretive Sociology*. Berkeley: University of California Press.

White, R. (ed.) (2010) *Global Environmental Harm: Criminological Perspectives*. Cullompton, Devon: Willan.

Wilkinson, D. (2002) *Environment and Law*. London and New York: Routledge.

Wolf, S. and Stanley, N. (2003) *On Environmental Law*. London: Cavendish.

Wood, M.C. (2009) 'Advancing the Sovereign Trust', *Environmental Law*, 39: 43–58.

6

Agribusiness, Governments and Food Crime: A Critical Perspective

Allison Gray and Ron Hinch

The relationships between humans and agricultural environments are not exempt from processes of modernization. Practices that involve industrialization, corporatization and neoliberalization have drastically reformed the modern practices of the food industry, as well as the regulations which govern them. Recognizing this context, Hazel Croall introduced the concept of food crime some years ago, defining it as the 'many crimes that are involved in the production, distribution and selling of basic foodstuffs' (2007: 206). Food crimes include a range of behaviours, ranging from economic and physical harms to both the humans and animals involved in the food industry, to food adulteration and the misrepresentation of food quality. Since then, while few other authors have specifically referred to their work as studies in food crime, there has been a growing literature dealing with connecting issues. As a focal point for criminological research, food crime overlaps with other emerging and established areas of inquiry, including green criminology (Beirne and South, 2007), environmental crime (White, 2008) and corporate crime (Croall, 2007).

Within critical criminology, food crime can be conceptualized as serious harms which need to be addressed beyond the traditional definitions of crime, allowing for the consideration of wider sociopolitical forces of harm (Beirne and South, 2007; White, 2007; Hil and Robertson, 2003). Within a harm perspective, an act or omission is not determined to be criminal by its illegal classification, but justice is extended beyond legality to harm recognition of negative consequences for a variety of both human and non-human victims (Benton, 2007). In the environmental crime field, the narrow focus on criminal definitions that reflect the interests of the powerful are opposed (Wright, 2011). Food crime is no different: food issues must be contextualized.

The relationship of food crime to the ideas of green criminology utilizes the medium of foodstuffs in understanding the fundamental associations between individuals and ecosystems. This leads into more complex relations between humans (farmers, consumers, food corporations, etc.), animals (livestock, wild, etc.), technologies (chemicals, machinery, biotech, etc.) and environments (agriculture, water, air, etc.), based on the safety and health of various concerns in sustenance, diet and nutrition. Many of these multifaceted relationships are organized with a multitude of policies, laws, regulations and rules, while many others, which arguably ought to be, are not. This (lack of) legal organization is a key area of analysis and criticism for researchers in food crime.

Given the need for a critical and holistic examination of food crime, and the importance of multiple actors in providing safe food, the concern here is to illustrate how the laws frequently fail to provide either safe working conditions for farmworkers, or the healthy food that consumers believe they are getting, while allowing agribusinesses to promote their interests. This will be illustrated through an examination of four of the more obvious examples of food crime involved in farming practices: farm labour in the cocoa industry, the impact of agribusiness on farming, pesticide use on agricultural crops and the impact of genetically modified organisms (GMOs) on farmer autonomy. These cases were chosen to allow for as broad an examination as possible while including hot-topic issues that have been adequately studied in academia. It is not the intent to suggest that these are the only areas of concern, but these examples will highlight the increasing power of agricultural industries through the corporatization of law in the food system.

It is important to note the use of the term 'government' throughout this chapter. Explaining the multitude of relationships, branches and details, both within and across governments, as well as connections to industries, is well beyond the scope of this chapter. In many cases the term 'governing bodies' has been used in order to broaden the definition to include all levels of government institution, both local and global. A detailed analysis of the fissures within the various food industries themselves is also omitted, but this should not be considered as unimportant. Both of these issues are welcomed as future research concerning this field.

Farm labour and low wages

Slavery has been part of human culture since ancient times (Heuman and Burnard, 2011), and its use in agricultural production is part of that history. This section will examine a portion of that history as it relates to the production of cocoa beans as an example of food crime.

From the time the Europeans stumbled over the Americas in 1492, the European economic exploitation of the region was dependent on slave labour, and the cocoa industry was no exception (Coe and Coe, 2007; Palmer, 1991). Slavery has been part of cocoa bean production ever since, and it has spread to other regions that were discovered as being able to grow cocoa beans. Initially, European colonizers enslaved the indigenous population, but it quickly proved to be problematic. This was because the introduction of European diseases to communities with no resistance to them led to significant population decline, as well as the unsuitability of indigenous labour to slave conditions, as perceived by Europeans. The solution was to turn to a previous and current source of slaves: Africa.

While there were those who resisted slavery, there were also many who advocated its use. It even received approval from the Pope (Panzer, 2008; Maxwell, 1975). The Catholic Church defended and promoted the slavery of non-Christian peoples. It was not until 1965 that the Church would definitively declare its opposition (Maxwell, 1975). Simultaneously, slavery debates were centred on English common law and a variety of legal cases and controversies (Ansley, 1975; Mtubani, 1983). Confusion and contradictions were rampant, with some courts refusing to recognize slavery while others permitted it (Mtubani, 1983). The profit-driven interests of the state became clear with the granting of a charter to the Royal African Company, monopolozing its control over the English slave trade, and later expanding its jurisdiction (Brown, 2007). It was not until 1840 that the buying, selling and ownership of slaves was banned in England (Brown, 2007).

While the battle over the formal status of slavery was being waged, the cocoa industry was expanding. The 1800s witnessed the beginnings of some of the world's most famous chocolate makers: Mars, Hershey, Cadbury, Rowntree, Fry and Lindt. It also witnessed the expansion of cocoa bean production from the Americas to other parts of the world. It is estimated that almost 55% of current cocoa production takes place in Africa, and that close to half of that occurs in Cote d'Ivoire (Coe and Coe, 2007).

As cocoa production grew, so did the movement to abolish slavery in the cocoa industry. One of the more dramatic events in that movement occurred in 1909, when George Cadbury, of the Cadbury Brothers chocolate fame, sued the *Standard* newspaper for libel, alleging that Cadbury Brothers had been wrongly accused in a *Standard* editorial of complicity in the slave trade (Satre, 2005). The Cadburys, a Quaker family that was associated with the Liberal Party, had been active in the anti-slavery movement of the time and alleged that this declaration by the *Standard*, associated with the Conservative Party, misrepresented the facts. The *Standard* editorial alleged that Cadbury was knowingly importing cocoa that was produced by slaves in the Portuguese colony of São Tomé.

It had been well established that slavery was being used in cocoa production on the island and the Portuguese government was doing nothing to stop it. The Cadbury family had undertaken its own investigation of the matter and concluded that while slavery was an issue, it was convinced that the Portuguese government was ready to move and abolish slavery in the colony and therefore took no action to find other, non-slave-produced, cocoa. The Portuguese government, however, did not act and slavery continued.

When the case went to trial, the jury concluded that the *Standard* had gone too far in its editorial, but it awarded the sum of just one farthing (a quarter of a penny). In such cases a trivial penalty is seen as a signal that while the jury believed that there was cause to say that the Cadbury company had been libelled, the libel was either trivial or Cadbury's reputation had been sullied by its own actions. It was not until 1975, after gaining independence, that slavery was formally abolished on São Tomé (Satre, 2005: 208). During this long battle to end slavery on the island, the Portuguese government consistently resisted abolition.

More recently, widespread concern was expressed in the worldwide media that slavery was alive and well after 2000 (International Labour Rights Forum, 2008), and there were increased efforts to put an end to it, especially as it involved the enslavement of children. In 2001, two US senators led a campaign to help to eliminate the use of unpaid child labour in the cocoa industry. Senators Harkin and Engel, as well as several anti-slavery groups, lobbied the big chocolate companies to join them in that effort. This produced the Harkin–Engel Protocol, which called for companies to devote significant resources towards reducing the use of forced child labour, establish an advisory committee to oversee implementation of the protocol by October 2001, eliminate the 'worst forms of child labour' (a euphemism for slave labour) and implement certification standards by 2005. The companies did create the advisory board, but it was disbanded in 2006 after the companies withdrew funding. The 2005 deadline for certification standards was extended to 2007, and extended again to 2010. Several commentators (Schrage and Ewing, 2005) have suggested that the protocol has resulted in reduced or eliminated child labour. Others are not so sure. The International Labour Rights Forum (2008) argues that while much has been accomplished to reduce slavery in the cocoa industry, much more remains to be done. Cocoa companies and governments alike have failed to provide the conditions that are needed to ensure the abolition of slavery.

Although most nations outlaw slavery and forbid the importation of slave-produced goods, these laws are generally poorly enforced. For example, the European Union (EU) passed a law in 2012 that requires all cocoa imported into Europe to not be produced by child slave labour, but it has been criticized for its weak enforcement provisions (Nieburg, 2012). Effectively none of the big multinational chocolate producers can say that their chocolate is produced without the use of unpaid and/or child labour. All major chocolate producers continue to resist independent verification that their cocoa supplies meet the criteria that are expressed in the Harkin–Engel Protocol (International Labour Rights Forum, 2008). The chocolate industry itself admits that under its own guidelines it will not be able to eliminate either unpaid or illegal child labour until 2020 (Peck, 2010).

The history of the legal status of slavery and the expansion of the cocoa industry remains a significant reason why the industry today continues to exploit populations that are involved in chocolate production. Although slave ownership has theoretically been banned throughout much of the modern world, cocoa labourers continue to experience exploitation, often mirroring the treatment of farm animals – a topic which environmental criminologists have consistently pointed to as deserving more attention, particularly in green criminology (Beirne, 1999; Cole, 2011; Sollund, 2008). As suggested in the Portuguese case, neither national governments nor dominant companies within the industry are keen to bring such issues towards solutions. Without effective laws and efficient enforcement of those laws by governing bodies, two consequences emerge. First, cocoa labourers will remain caught in their slave-like conditions, forced to work in environmentally unsafe conditions (see Off, 2006), and second, chocolate companies will continue to profit from such an unjust environment.

Agribusiness vs. 'traditional' farming

Prior to the introduction of modern agricultural systems, food production involved a farming system that was a completely self-contained industry. This 'traditional' farming meant that all inputs, processes and products that were necessary for food production were dealt with by individual farmers. Throughout the Industrial Revolution a variety of technological innovations, together with the scarcity of farm labour, pushed the development of input agribusiness, where labour was replaced by machines, production increased, and farmers began purchasing inputs from off-farm locations (Woolverton et al., 1985). Agribusiness, as defined by Davis and Goldberg (1957), is the 'sum total of all operations involved in the manufacture and distribution of farm supplies' (1). The blossoming of the agribusiness industry led to increased food production in response to a growing population, and the desires to process, manufacture and distribute that food were placed into agribusiness jurisdiction (Woolverton et al., 1985).

The study of agribusiness requires the inclusion of all components which are impacted by its behaviour. These include a variety of actors, including farmers, governmental agencies, consumers and so on. The focus here is on the changing nature of agribusiness and the first two: farmers and governing bodies. Their relationship within the modern industry is complex and unique, as Thompson (1996) argues that the large amount of inputs, over extended biological production stages, makes agriculture a very capital-intensive industry. During the late 20th century, agribusiness became characterized by increasing complexity (global networks), concentration (vertical integration and large investors' financial power), competence (management of increasing technology and production) and competitiveness (demand-driven and increasing numbers of suppliers), yet was compromised (via the minority position of farmers in the agricultural world) (French, 1989). However, agribusiness is understood as being a key factor in integrating local farmers into global markets (Echánove and Steffen, 2005).

Several key events showcase the adverse impact of modern (global) agriculture on local farmers. This became very clear during the 1980s financial crisis, when farmers suffered from a lack of access to loans and there was an increase in the number who filed for bankruptcy (Ginder et al., 1985). One significant process that has enabled the growth in power of modern agriculture over traditional farming is the development of contract farming. Agribusiness, as an industry, prefers contractual relationships since these eliminate moral considerations such as performance-based wages (Ogishi et al., 2003). Further, various laws changed the legal definitions and treatment of farmers. For example, prior to the mid-1980s, farmers were treated better within US income tax law: they could deduct production inputs from sales receipts in the year of purchase rather than in the year of use, and depreciate their capital assets at a faster rate than the useful lives of those assets (Thompson, 1996). Over time, these changes have helped to progress the (financial) power of large agribusiness companies at the expense of local farmers.

This negative relation between the emerging agribusiness industry and farmers takes place on both a local and a global scale. The international consolidation of agribusiness, particularly through an oligopoly of US and EU companies, leaves developing nations in a subordinate position. As Wilkinson (2009) argues, even developing nations such as Brazil and China, which have been able to preserve domestic (traditional) agriculture so far, do not have a positive-looking future concerning their competitiveness in the global agricultural industry. Within local agriculture there are similar power relations, where large agribusiness corporations are prioritized over local farming. Due to financial pressures within this competitive industry, farmers are increasingly becoming the 'rural landless': they live on the land but lack decisive power in what is grown and for whom, resulting in an emotional divorce from the land (Tandon, 2010). This subordination has resulted in farmers being the weakest element in the food industry (Buccirossi et al., 2002). The overall relationship is filled with conflicts: national versus global, small versus big, agricorporations versus family farmers (French, 1989), including a prioritization of the global 'side', where export agriculture is valued over the local farming efforts which feed domestic populations (Tandon, 2010).

National governments can play an important role in mediating the agribusiness shift and its impact on local food production. Their involvement in the food industry is necessary for several reasons, including global trade, food safety and adequate infrastructure. For example, public agencies need to provide the transport facilities that will enable the flow of inputs and products (Thompson, 1996). There is also growing concern for environmental impact, and without government regulation of the food industry, agribusiness corporations would underinvest in pollution control – something which market forces cannot singularly regulate (Schroeter et al., 2006). The key purpose of government intervention is the management of private land ownership, where domestically there are some government initiatives which exclude farmers with lower acreage or production levels from the opportunities that are given to agricorporations, amid a growing global agridomination in the developing world through land-grabbing (Tandon, 2010). Although there is a clear need for government intervention, the problem occurs when private corporate interests have a say in the laws and regulations which govern them.

For instance, there are agricultural laws which aim to control the structure of the farming industry. In the US cattle feed industry, some states have 'anti-corporate' farming laws which restrict corporations from intervening in food production. These laws are meant as a protective measure to preserve the family farm and thus the ownership of agriculture, but they also indirectly impact the size structure of the industry. Schroeter et al. (2006) performed a cross-comparison between the top four cattle-feed producing states, where one – Nebraska – held such anti-corporate farming laws, but they found no strong evidence of these regulations affecting the cattle-feed industry. However, across all four states they did find an overall steady trend of feedlots increasing in size, suggesting that agribusiness slowly creeps into an industry regardless of the agricultural laws in place.

At the other end of this process is the byproduct of farmed livestock: animal waste. In the USA, pollution from animal waste is regulated largely through technological standards, subsidization and non-compliance fines, all of which are ineffective due to differences in responsibility or liability for animal waste pollution (Ogishi et al., 2003). When animal waste pollution is discovered, only the producers who raise the livestock are responsible for the environmental damage which ensues. Corporations, outside facility ownership, which own the animals, are not accountable for the waste or the damage done. Due to the changes in agribusiness such as increasing (vertical) coordination and contract farming, it is becoming difficult to hold producers responsible for their damage to the environment (Ogishi et al., 2003).

These two cases exemplify the inadequacy of laws to create and maintain healthy food and a safe environment. The cattle feedlot industry shows that laws do not stop agribusinesses from getting their way, and the animal waste example shows that laws are not actually effective, while letting agribusinesses have minimum responsibility for their behaviour at the expense of producers. In addition, the following case shows how governing bodies move away from this passive stance shown in the previous cases and take an active role in promoting the interests of agribusiness. In Mexico, farmers feel pressurized into entering into contracts with export-oriented agribusinesses because they lack alternative financing or technological assistance, while having limited secure market access individually (Echánove and Steffen, 2005). Contract farming is accompanied by neoliberal policies which further disadvantage individual farmers. For instance, the government wanted to promote grain production in 1998 so it offered extra money to those producers who would grow corn or wheat under contract with agribusiness (Echánove and Steffen, 2005). In effect, individual farmers are subordinated within the industry, being told what to grow, how much and to whom to sell it.

Beyond the crimes, harms and injustice that are experienced by animals within many agribusiness livestock complexes, this section has focused on how agribusinesses, with control and power as their focal interests, are not only surpassing the law, but are encouraging and enabling their interests through partnerships with governing bodies. Farmers are directly impacted through the subordination of their labour through contract farming, and many who do not secure such markets find themselves at the brink of bankruptcy. Given the growth of international agribusiness corporations and the increasing neoliberal policies of local governments, there is little hope that this situation will change for the better in the near future.

Regulating pesticide use on agricultural crops

The use of dichlorodiphenyltrichloroethane (DDT), an organochlorine (OC) insecticide, began as a method to combat malaria (Smith, 2000). It was sprayed on homes to safeguard against virus-carrying mosquitos entering the home (Attaran et al., 2000). This was, and still is in some parts of the world, a very effective method. It was not until the late 1940s that DDT began to be used as an agricultural pesticide (insecticide), becoming widely adopted until the 1970s (Stemmler and Lammel, 2009), with extensive use in agriculture in the USA from the 1940s to the 1960s (Longnecker et al., 1997).

The movement to ban DDT is largely thought of as an ecological one, when research found problems in the reproductive processes of many species of birds that were exposed to the chemical. It should be noted that DDT is theoretically not banned globally, and that any country may use the chemical; the global ban failed when many countries and scientists lobbied for its continued use against malaria (Mandavalli, 2006). However, the movement to ban it signalled a substantial reduction in the environmental harm that is inflicted via pesticide use (Angelo, 2008).

To take the place of DDT and other OCs, farmers switched to organophosphates (OPs) on the advice of governing bodies and pesticide producers because these were said to be a safer alternative. The change was justified in three ways. First was to minimize the adverse health effects on people who are directly exposed to DDT. Second was that OCs accumulated in the environment much more than OPs, having a greater impact on the toxicity levels in water sources, soils and animals (Stemmler and Lammel, 2009; Rogan and Chen, 2005). Third was the governmental portrayal of OPs as safer alternatives to DDT and other OCs, coupled with the OP manufacturers taking advantage of the DDT ban attempts and promoting the use of OPs as if they were a safer option (Angelo, 2008). These will each be explored below.

The impact of OCs on human health was arguably the least important factor in the anticipation of the DDT ban, but was still a key consideration. DDT exposure is linked to a variety of human health concerns, including cancers, reproductive issues and neurological disorders (Longnecker et al., 1997). The switch to OPs did not solve these health problems because these are just as toxic to human health (see Goodman, 2011 for research on pesticide exposure in the womb): recall that varieties of OPs can be traced back to being used as forms of nerve gas in the Second World War. Obviously, human health is threatened most significantly among the farmers who apply the pesticides (Murphy-Greene and Leip, 2002), but the switch to OPs is not a solution to the adverse health effects of pesticide exposure. Current information from the Center for Environmental Research and Children's Health (2012) claims that exposure to large amounts of OP pesticides is more harmful to human health than the same amount of OC pesticides.

The impact of pesticides on the environment was a second key reason for switching from OCs to OPs. However, not only are OPs also very harmful to the environment but the laws that were created to regulate these pesticides are themselves ill-defined and poorly enforced, causing and perpetuating environmental harms (Angelo, 2008). After the switch to OPs there were a record number of wildlife poisonings that were attributed to drifting spray (Rattner, 2009). There are even suggestions that OPs are included in the EU Black List due to their toxicity, especially to the aquatic environment. However, evidence of this cannot be confirmed due to corporate financial fears about product aversion (see Surman, 2009). The laws governing OP use are also poorly enforced. For example, the regulations specify that OP pesticides are to be used only when birds are not breeding. However, one data source suggests that there were more than 4,000 incidents of pesticide poisoning in the USA resulting from the inappropriate use of OPs and that these incidents contributed to the death of more than 40,000 birds (Angelo, 2008).

At first glance the efforts to ban DDT and to limit the use of OPs (through avoiding bird breeding seasons) seem well reasoned and should protect human health, wildlife health and ecosystems. However, given the complex relationship between humans and the environment, including human problems and environmental problems being intrinsically connected, there is reason to suggest that there is some concern about an overemphasis of an anthropocentric perspective accompanying pesticide regulation. Benton (2007) defines an anthropocentric orientation as 'recognizing and valuing the ecocentric orientation to the world as a human purpose' (26). The research on the environmental impact of pesticides has shown that the switch from OCs to OPs has not benefited the environment: even the idea that OPs have lower bioaccumulation levels has been targeted via new research that shows the presence of such pesticides in soil and drinking water long after their application (see Nigam et al., 2011; Mugni et al., 2012).

However, the connection between environments and the food chain was a third concern prompting the switch to OPs: namely, the high bioaccumulation of OC pesticides in organisms (including humans) and along the food chain (Stemmler and Lammel, 2009; Rogan and Chen, 2005). The similarities of both types of pesticide in their impact on human health with exposure (as well as causing negative effects on the environment) suggests that the main reason for the switch was to decrease the bioaccumulation of toxins in the food chain. If this is the case, then harm has been defined through human-centred values of protecting the food chain as it concerns human health. OP manufacturers were able to lobby for the safety of their products during the movement to ban DDT, arguably driven by profit rather than a concern for the environment, which still suffers the consequences of harmful pesticides.

Beyond this anthropocentric ideology, laws that govern human, animal and environmental health often overlap significantly, translating to several different acts, agencies and regulations which come into play simultaneously and sometimes in a contradictory manner. This confusion can create loopholes or uncertainties in the laws, limiting the feasibility of following the laws and enforcing them. For example, the Endangered Species Act and the Federal Insecticide, Fungicide and Rodenticide Act are both designed to provide environmental protection against pesticide use, but both are ill-defined with weak enforcement provisions. In particular, the laws fail to specify the type of and process for consultation with associated agencies (Angelo, 2008).

Human health enforcement fairs no better. Policies that aim to protect farmworkers from hazardous pesticide materials failed due to a lack of effective monitoring of how much pesticide was actually being used and how much exposure workers experienced (Murphy-Greene and Leip, 2002). This, along with inadequate funding and a lack of inspectors, means that pesticide laws are often inadequately enforced and thus fail. For example, there were cases in Florida where a large number of Spanish-speaking farmworkers had to use products which contained written warnings about their hazardous nature that were only in English – Spanish warnings were unavailable (Murphy-Greene and Leip, 2002).

These cases exemplify how the laws and regulations are inadequate in protecting farmers, animals and the environment from the harmful effects of pesticide use and exposure. Many laws that govern pesticide use are either poorly defined or fail to achieve their enforcement objectives when governments do not provide the necessary resources. This phenomenon allows agribusiness companies that manufacture the pesticides to continually harm the environment and put farm labourers at risk of serious health problems while increasing their profitable interests.

Patented seeds and farmer autonomy

One of the most controversial areas of food production is the creation of patent laws that allow agritech businesses to effectively patent life forms. Genetically modified foods (GMFs), part of GMOs, are part of the hotly debated patenting issue, particularly concerning the morality behind producing and consuming crops from seeds that are engineered to contain selective elements of other species or bacteria in order to resist certain diseases, herbicides or pests. Patent laws have been around for a long time, dating back almost 150 years (Lerner, 2002), but as noted by Simmons (2013), a large number of patents have been granted for food products since 2000 (in the USA, 42% of all GMF patents were granted between 2000 and 2008). Globally the majority of GM seeds (and their accompanying fertilizers and herbicides) are produced from four main corporations – Monsanto, Syngenta, Du Pont and Bayer (Nottingham, 2003) – suggesting a form of bioimperialism (Engdahl, 2004).

However, a seed patent in food crops means that the seed can only be used once, during one growing season, and the farmer growing that plant must enter into a strict contract with the producing company (which often includes provisions concerning to whom the farmer can sell the crop). The farmer is not able to plant that seed without entering into a contract, nor can they save seed from the season's harvest and use it to plant the following season's crop. This does not seem far-fetched, with hybrid and organic seeds falling into similar patenting issues. However, given the natural atmosphere of farming and the openness of farm fields in the environment, it becomes easy to see how seed does not necessarily always stay where it is planted.

Thus when seeds blown into a non-contract field grow, the farmer has to have a licence/contract for even a single plant or risk being fined. Monsanto Canada's director of public affairs has stated: 'it is not, nor has it ever been Monsanto Canada's policy to enforce its patent on Roundup Ready crops when they are present on a farmer's field by accident... Only when there has been a knowing and deliberate violation of its patent rights will Monsanto act' (Schubert, 2002: para. 8). So although the agricorporations do not physically own the genetically modified plants, the patent laws involving genetic-alteration processes are utilized to serve the interests of the corporation. Accordingly, it is not the intent to debate the ethics of GMFs here, but to analyse the specific consequences of allowing these patents. More specifically, the focus is on analysing the loss of control that farmers and communities experience as a result of the (direct or indirect) use of patented seeds.

As might be expected, farmers have launched a large number of lawsuits aimed at resisting the GMF patents of agritech firms, some of which are pre-emptive lawsuits to avoid patent infringement charges (see Fairley, 1999; Boone, 2013; Fatka, 2013). Also not to be unexpected, agritech businesses have launched legal actions against farmers for the unauthorized use of patented GM seeds. Some lawsuits are against farmers who innocently grew GMFs (Walters, 2007). A few of the better-known court cases include *Monsanto Canada vs. Schmeiser* in 2004 (see Wiber, 2009), *Hoffman vs. Monsanto Canada* in 2005 (see Judge, 2007) and *Bowman vs. Monsanto Co. et al.* in 2013 (see Simmons, 2013). In each of these cases the central issue concerns the ability of the farmer to use traditional farming methods while knowingly using patented seeds.

For example, in the Schmeiser case, the farmer sought to defend the traditional practice of saving a portion of the crop that is grown each year to use as seeds for the following growing season. He had not purchased the patented Monsanto canola seed and thus was not a licensed grower, but he obtained seeds after noticing that some of his crop, particularly along the roadways next to fields where licensed canola was grown by other farmers, was resistant to his weedkiller (Wiber, 2009). He harvested the resistant canola seed and planted it the following year. In subsequent court action it was ruled that he had violated Monsanto's patent by deliberately using what he knew to be patented seed. His defence that it was windblown seed and that he had a traditional right to plant windblown seed was rejected by the court. However, the court upheld Monsanto's patent rights, saying that these rights negated farmers' property rights to the seeds found on their farms. Similar rulings were made in the other cases.

This outcome of the war between traditional farmer property rights and the newer GM company patent rights has adverse consequences for farmer autonomy. As previously discussed, the power relations between farmers and agribusinesses are considered to disproportionately favour the latter, which is largely due to the importance of the economic buying power of food processors and manufacturers, especially given the perishable nature of food (Carolan, 2012). Toews (2008) takes a critical stance regarding the issue of farmer control, suggesting that farmers are a *petit bourgeois* – a social group that is caught somewhere between those with power and those without, while believing that they ultimately control their farming methods. As exemplified in the outcome of the Schmeiser case, this 'illusion of independence' may not continue into the near future because of the significant involvement of the state via the court system. Not only did the courts in each case reaffirm the patent protections that had been granted to Monsanto, but they also extended the prioritization of patent laws to other multinational agritech industries.

Looking beyond the immediate concern of these individual farmers, the impact of these decisions to enforce the patent laws is far reaching. Farmers the world over rely on being able to do as Schmeiser did – retain a percentage of each year's crop to plant the following season. This practice reduces the seed costs in the next season and gives the farmer more control over what is planted. Farmers have argued that the ability to save seeds reduces their costs and allows them to continue farming. Without this ability to use their seeds as they so choose, including planting of patented seed that is blown onto their farms from neighbouring farms, farmers argue that they become contract farmers and lose the ability to make independent decisions regarding their farming (Wiber, 2009).

Finally, it must be pointed out that the various patents are created in a context in which agritech industries lobby governments for the protection of their products and thus their patents. As Lerner (2002) and Ginarte and Park (1997) have suggested, this often results in conflicts between agritech industries located in different countries.

The governments in these countries seek to enhance the viability of their native industries, giving these preferential treatment. One such example is the fight over GMOs, such as Monsanto's patented canola, which has pitted North American and European interests against each other in the battle to control European marketing (Morgan and Goh, 2004). US agritech companies have fought vigorously, with the aid of the US government in particular, before the Word Trade Organization (WTO) to gain access to European markets (see Monsanto's case in Walters, 2006). In the process, governments have become the champions of the agritech industries within their borders, while divisions are constructed and maintained through the differences in GM laws and regulations globally – a thought which leads to major concerns about international regulation and enforcement.

It would be a mistake, however, to believe that GMF agritech companies have it all their own way. Some nations can, and have, said no, such as Zambia's refusal to admit GMFs into their ecosystem. In a war of communication, Western leaders (including GM corporations) argued that millions were starving in Zambia, and the introduction of GMFs would eliminate such food shortages. However, according to local populations, millions of people were not starving, and such claims ignored the unequal access to food and the concentrated nature of the local food production system across Zambia (Walters, 2006). There are other nations which have successfully battled GM corporations, including Monsanto being forced to pay $1.5 million fines for attempting to corrupt government officials in an effort to gain access to the Indonesian market from 1997 to 2002 (Walters, 2006). This type of pressure on countries to accept GMOs is a clear example of how GM corporations (together with their national states) exploit international environmental law for political and economic gain.

To conclude, agritech corporations have employed a variety of legal and sometimes illegal methods to promote their business interests. They have used patent laws to overturn traditional farming practices, sometimes penalizing even those farmers who have innocently grown GMFs. They have collaborated with governing bodies, including national states and international organizations such as the WTO, whenever they encounter resistance to their products. Thus patent laws are not only poorly framed, virtually allowing the control of life forms and not just the GM process, but they are also poorly enforced, as exemplified by the successful litigation against farmers who innocently grow GMFs, such as Schmieser. Together with their local governing bodies,

GM manufacturers push for these patent laws, allowing their interests to dominate while seizing more control at the expense of farmer autonomy.

Conclusion

The intent of this chapter was to illustrate ways in which governing bodies and agritech industries have consistently worked together over time to promote the interests of the latter through laws and regulations. These laws provide significantly better protection of the concerns of agritech industries while providing minimal or ineffective support for the protection of other interests, including the environment, cultural tradition, and farmers or farmworkers. The case studies analysed showcase the various forms of agribusiness and agritech companies that use and overpower the law and its enforcement in order to both promote profit-oriented goals and increase their power or control over the food industry: from traditional farming culture, where local farmers could save their seed and were able to farm efficiently without toxic pesticides within self-contained production methods, to powerful agribusinesses, which wage legal battles over seed patents, while working to have their products utilized regardless of harming the environment and allowing the continued exploitation of farm labourers in slave-like conditions. Not only are the current laws (if existing) ineffective in providing safe environments, healthy food and healthy people but they are being (re)constructed, neglected and overridden by agritech corporations and agribusinesses, often in collaboration with both local and global governing bodies.

Acknowledgement

We would like to express our thanks to Steve Tombs for his insightful comments on previous drafts of this chapter.

References

Angelo, M. (2008) 'The killing fields: Reducing the casualties in the battle between U.S. species protection law and pesticide law', *The Harvard Environmental Law Review*, 32(1): 95–148.

Ansley, R. (1975). *The Atlantic Slave Trade and British Abolition.* London: Macmillan.

Attaran, A., Maharaj, R. and Liroff, R. (2000) 'Doctoring malaria, badly: The global campaign to ban DDT', *British Medical Journal, International Edition*, 321(7273); 1403–1405.

Beirne, P. (1999) 'For a nonspeciesist criminology: Animal abuse as an object of study', *Criminology,* 37(1): 117–147.

Beirne, P. and South, N. (eds) (2007) *Issues in Green Criminology: Confronting Harms against Environments, Human and Other Animals.* Portland, OR: Willan Publishing.

Benton, T. (2007) 'Ecology, community and justice'. In Beirne, P. and South, N. (eds) *Issues in Green Criminology: Confronting Harms against Environments, Human and Other Animals.* Portland, OR: Willan Publishing, pp. 3–31.

Boone, R. (2013) 'GMO wheat lawsuit: Idaho farmers sue Monsanto', *Huffington Post.* Accessed on 26 May 2014 at http://www.huffingtonpost.com/2013/06/12/gmo-wheat-lawsuit-idaho_n_3430961.html.

Brown, C. (2007) 'The British government and the slave trade: Early parliamentary enquiries'. In Farrell, S. Unwin, M. and Walvin, J. (eds) *British Slave Trade: Abolition, Parliament and People.* Edinburgh: Edinburgh University Press, pp. 27–41,

Buccirossi, P., Marette, S. and Shiavina, A. (2002) 'Competition policy and the agribusiness sector in the European union,' *European Review of Agricultural Economics,* 29(3): 373–397.

Carolan, M. (2012) *The Sociology of Food and Agriculture.* New York: Routledge.

Center for Environmental Research and Children's Health (2012) *Organophosphate Pesticides.* Accessed on 26 May 2014 at http://cerch.org/environmental-exposures/organophosphate-pesticides/.

Coe, S. and Coe, M. (2007) *The True History of Chocolate,* 2nd edn. London: Thames and Hudson Books.

Cole, M. (2011) 'From "animal machines" to "happy meat"? Foucault's ideas of disciplinary and pastoral power applied to "animal-centred" welfare discourse', *Animals,* 1(1): 83–101.

Croall, H. (2007) 'Food crime'. In Beirne, P. and South, N. (eds) *Issues in Green Criminology: Confronting Harms against Environments, Human and Other Animals.* Portland, OR: Willan Publishing, pp. 206–229.

Davis J. and Goldberg, R. (1957) *A Concept of Agribusiness.* Boston: Harvard University.

Echánove, F. and Steffen, C. (2005) 'Agribusiness and farmers in Mexico: The importance of contractual relations', *The Geographical Journal,* 171(2): 166–176.

Engdahl, F. (2004) 'Bio-imperialism: Why the biotech bullies must be stopped', In *Gene-Manipulated Seeds: Are We Losing Our Food Security Too?.* Accessed on 26 May 2014 at http://www.organicconsumers.org/monsanto/muststop090304.cfm.

Fairley, P. (1999) 'Farmers sue seed firms', *Chemical Week,* 35: 13.

Fatka, J. (2013) 'Organic group sues over GM patents: Organic groups attempt to pre-emptively Sue Monsanto to protect from being accused of patent infringement', *Feedstuffs,* 85(38): 23.

French, C. (1989) 'The changing face of agribusiness', *Agribusiness,* 5(3): 217–227.

Ginarte, J.C. and Park, W.G. (1997) 'Determinants of patent rights: A cross-national study, *Research Policy,* 26(3): 283–301.

Ginder, R., Stone, K. and Otto, D. (1985) 'Impact of the farm financial crisis on agribusiness firms and rural communities', *American Journal of Agricultural Economics,* 67(5): 1184–1190.

Goodman, B. (2011) 'Pesticide exposure in womb linked to lower IQ'. Accessed on 28 May 2014 at http://www.webmd.com/baby/news/20110421/pesticide-exposure-in-womb-linked-to-lower-iq.

Heuman, G. and Burnard, T. (eds) (2011). *The Routledge History of Slavery*. New York: Routledge.

International Labour Rights Forum (2008) 'The cocoa protocol: Success or failure'. Accessed 26 May 2014 at http://www.laborrights.org/sites/default/files/publications-and-resources/Cocoa%20Protocol%20Success%20or%20Failure%20June%202008.pdf.

Hil, R. and Robertson, R. (2003) 'What sort of future for critical criminology?' *Crime, Law and Social Change*, 39(1): 91–115.

Judge, E. (2007) 'Intellectual property law as an internal limit on intellectual property rights and autonomous source of liability for intellectual property owners', *Bulletin of Science Technology Society*, 27(4): 301–313.

Lerner, J. (2002) '150 years of patent protection', *The American Economic Review*, 92(2): 221–225.

Longnecker, M., Rogan, W. and Lucier, G. (1997) 'The human health effects of DDT (dichlorodiphenyl-trichloroethane) and PCBS (polychlorinated biphenyls) and an Overview of Organochlorines in public health', *Annual Review of Public Health*, 18(1): 211–244.

Mandavalli, A. (2006) 'Health agency backs use of DDT against malaria', *Nature*, 443(1): 250–251.

Maxwell, J. (1975) *Slavery and the Catholic Church: The History of the Catholic Teaching Concerning the Moral Legitimacy of the Institution of Slavery*. London: Barry Rose Publishers. Accessed at from http://www.anthonyflood.com/maxwellslavery01.PDF.

Morgan, D. and Goh, G. (2004) 'Genetically modified food labelling and the WTO agreements', *Review of European Community & International Environmental Law*, 13(3): 306–319.

Mtubani, V. (1983) 'African slaves and English law', *Journal of African Studies*, 3(2): 71–75.

Mugni, H., Demetrio, P., Paracampo, A., Pardi, M., Bulus, G. and Bonetto, C. (2012) 'Toxicity persistence in runoff water and soil in experimental soybean plots following chlorpyrifos application', *Bulletin of Environmental Contamination and Toxicology*, 89(1): 208–212.

Murphy-Greene, C. and Leip, L. (2002) 'Assessing the effectiveness of executive order 12898: Environmental justice for all', *Public Administration Review*, 62(6): 679–687.

Nieburg, O. (2012) 'EU resolution on cocoa child labour has "no bite" says labour group'. *Copnfectinarynew.com*. Accessed 3 September 2013 at http://www.confectionerynews.com/Commodities/EU-resolution-on-cocoa-child-labour-has-no-bite-says-Labour-group.

Nigam, S., Singh, J., Luxmi Singh, A., Das, V. and Singh, P. (2011) 'Organochlorines, organophosphate bioaccumulation and reproductive dysfunction in fish captured from polluted river Gomti during pre-monsoon', *Journal of Ecophysiology & Occupational Health*, 11(1/2): 9–19.

Nottingham, S. (2003) *Eat Your Genes: How Genetically Modified Food Is Entering Your Diet*. New York: Zed Books Ltd.

Off, C. (2006) *Bitter Chocolate: Investigating the Dark Side of the World's Most Seductive Sweet*. Toronto: Random House.

Ogishi, A., Zilberman, D. and Metcalfe, M. (2003) 'Integrated agribusiness and liability for animal waste', *Environmental Science & Policy*, 6(2): 181–188.

Palmer, C. (1991) 'The early black diaspora in the Americas: The first century after Columbus', *OAH Magazine of History*, 5(4): 27–30.

Panzer, J. (2008) 'The popes and slavery', *The ChurchinHistory Information Centre*. Accessed on 26 May 2014 at http://www.churchinhistory.org/pages/booklets/slavery.pdf.

Peck, S. (2010) 'Certifying blood chocolate', *In These Time*. Accessed on 26 May 2014 at http://inthesetimes.com/article/6108/certifying_blood_chocolate.

Rattner, B. (2009) 'History of wildlife toxicology', *Ecotoxicology*, 18(7): 773–783.

Rogan, W. and Chen, A. (2005) 'Health risks and benefits of bis(4-chlorophenyl)-1,1,1-trichloroethane (DDT)', *The Lancet*, 366(9487): 763–773.

Satre, L. (2005) *Chocolate on Trial: Slavery and the Ethics of Business*. Athens: Ohio University Press.

Schrage, E. and Ewing, A. (2005) 'The cocoa industry and child labour', *The Journal of Corporate Citizenship*, 18: 99–112.

Schroeter, R., Azzam, A. and Aiken, D. (2006) 'Anti-corporate farming laws and industry structure: The case of cattle feeding', *American Journal of Agricultural Economics*, 88(4): 1000–1014.

Schubert, R. (2002) *'Schmeiser Wants to Take It to the Supreme Court'*. *Cropchoice.com*. Accessed on 26 May 2014 at http://www.cropchoice.com/leadstryb5ea.html?recid=935.

Simmons, W. (2013) 'Bowman v. Monsanto and the protection of patented replicative biologic technologies', *Nature Biotechnology*, 31(7): 602–606.

Smith, A. (2000) 'How toxic is DDT?' *The Lancet*, 356(9226): 267–268.

Sollund, R. (2008) 'Causes for speciesism: Difference, distance and Denia', In Sollund, R. (ed.) *Global Harms: Ecological Crime and Speciesism*. New York: Nova Science Publishers, pp. 109–130.

Stemmler, I. and Lammel, G. (2009) 'Cycling of DDT in the global environment 1950–2002: World ocean returns the pollutant', *Geophysical Research Letters*, 26(24): 24602.

Surman, W. (2009) *'Industry denies pesticide black-list'*, *Farmers Guardian*. Accessed on 26 May 2014 at http://www.farmersguardian.com/industry-denies-pesticide-black-list/23663.article.

Tandon, N. (2010) 'New agribusiness investments mean wholesale sell-out for women farmers', *Gender & Development*, 18(3): 503–514.

Thompson, R. (1996) 'Impact of budget and tax policy on agriculture and agribusiness: The American experience', *Agribusiness*,12(6): 601–611.

Toews, R. (2008) 'Petit bourgeois consciousness and farmers in Western Canada: A re-assessment of CB MacPherson', *English Quarterly*, 40(1/2): 20–27.

Walters, R. (2006) 'Crime, bio-agriculture and the exploitation of hunger', *The British Journal of Criminology*, 46(1): 26–45.

Walters, R. (2007) 'Food crime, regulation and the biotech harvest', *European Journal of Criminology*, 4(2): 217–235.

White, R. (2007) 'Green criminology and the pursuit of social and ecological justice'. In Beirne, P. and South, N. (eds) *Issues in Green Criminology: Confronting*

Harms against Environments, Human and Other Animals. Portland, OR: Willan Publishing, pp. 32–54.

White, R. (2008) *Crimes against Nature*. Cullompton: William Publishing.

Wiber, M. (2009) 'What innocent bystanders? The impact of law and economics reasoning on rural property rights', *Anthropologica*, 51(1): 29–38.

Wilkinson, J. (2009) 'Globalization of agribusiness and developing world food systems', *Monthly Review: An Independent Socialist Magazine*, 61(4): 1–9.

Woolverton, M., Cramer, G. and Hammonds, T. (1985) 'Agribusiness: What is it all about?' *Agribusiness*, 1(1): 1–3.

Wright, G. (2011) 'Conceptualising and combating transnational environmental crime', *Trends in Organized Crime*, 14(4): 332–346.

7

Anthropogenic Development Drives Species to Be Endangered: Capitalism and the Decline of Species

Michael J. Lynch, Michael A. Long and Paul B. Stretesky

Introduction

The study of crimes against non-human species is central to the development of green criminology. This discussion contributes to this particular area of green criminology by examining limits to the viability of non-human animal populations as a function of systemic ecological harms.[1] By 'systemic' we mean those ecological harms that are endemic to capitalism as a system of production, and which in the current context of global capitalism are also global in their appearance, and therefore are structural in origin. The specific viability issue that we address is the endangerment and extinction of species, and the relationship of species' viability to the forms of ecological disorganization that are produced by capitalism in its ordinary course of development. We examine these issues in relation to the tendency for capitalism to produce ecological disorganization.

One reason criminologists should study ecological disorganization is that it draws attention to violations of human rules of law and/or natures' rules of law related to physics, chemistry and planetary boundaries (Long et al., 2014; see generally Carson, 2002; Colborn et al., 1996; Ehrlich and Ehrlich, 1996; Foster et al., 2010; Steingraber, 1997; Rockstron et al., 2009; Wargo, 1998). As a result, green criminologists may illustrate how social and ecological harms can expand criminology, as a science, by rejecting the state definition of crime as the only valid method for examining crime (Long et al., 2014). Thus criminologists can examine crime as a function of ecological organization and the normal functioning of ecosystems (Stretesky et al., 2013; Lynch et al., 2013). As part of those efforts to analyse and understand green crimes, green criminological studies have drawn significant attention to crimes against non-human species (Aatola, 2012; Beirne, 1999, 2009; Bjørkdahl, 2012; Clarke and Rolf, 2013; Eliason, 2012; Hagstedt and Korsell, 2012; Ngoc and Wyatt, 2013; Nurse, 2013; Pires and Clarke, 2012; Sollund, 2008, 2011, 2012a, 2013; Svärd, 2012; Ween, 2012; Wyatt, 2009, 2011). In particular, some green criminological studies illustrate the impact of these green crimes by focusing specific attention on activities that have negative non-human species impacts through illegal activities such as poaching, hunting and the illegal trade in wildlife (e.g. Clarke and Rolf, 2013; Hagstedt and Korsell, 2012; Pires and Clarke, 2012; Pires and Moreto, 2011; Sollund, 2011; Wyatt, 2013). Studies of non-human animal harms have significantly expanded the concept of (green) victimization beyond the traditional criminological limits, which draws attention only to human victims.

There is much to be learned from case studies of poaching, the illegal trade in wildlife and the impacts of these activities, and the in-depth studies that green criminologists have carried out on these issues have certainly expanded our knowledge of the details of how many crimes against non-human animals unfold, their scope and the specific kinds of victimization that they involve. At the same time, the existing green criminological literature on non-human species victimization has yet to develop a thorough analysis of the broader problems that non-human species face, such as species decline in relation to the structural causes of those outcomes. The structural factors that are related to species decline are broad and include processes such as deforestation, habitat loss and climate change, which impact non-human species (see Sollund, 2013), including lesser studied entities such as insects, fungi, molluscs and algae that play important roles in maintaining ecosystem functionality and balance. In our view, one of the structural explanations that deserve greater attention is the way in which the capitalist treadmill of production generates widespread non-human species victimization (Stretesky et al., 2013; Lynch et al., 2013). The treadmill of production is a framework that was proposed by Alan Schnaiberg (1980; see also Gould et al., 2008; Schnaiberg and Gould, 1994). It helps to explain how the capitalist system accelerates ecological disorganization by requiring more production in an effort to continuously grow and accumulate profit. From the view of treadmill theorists, environmental problems – including the extinction and decline of non-humans – will grow as long as the current political economic conditions continue to exist (see also Foster et al., 2010). While there are exceptions (e.g. Pellow, 2004; Stretesky et al., 2013; Vail, 2009), green criminologists have not yet seriously entertained structural analysis that is specifically embedded in a treadmill of production framework as a mechanism for explaining crimes/harms against non-human species. We draw attention to this issue here.

To illustrate the effects of human economic development on the more general decline of wildlife species,[2] we first review how the contemporary expansion of capitalism drives continuous economic development in ways that promote the destruction of nature and facilitate the general decline of species health, vitality and existence. In addition, to support this argument we make reference to scientific literature on the Anthropocene extinction and current elevated rates of extinction to demonstrate the broad scope of extinction in the contemporary era that would be consistent with taking a structural view of these kinds of negative outcome for non-human species. We then illustrate the more general state of species endangerment through reviews of the International Union for the Conservation of Nature's (IUCN) Red List of threatened species, and data on species endangerment from the US Fish and Wildlife Service (USFWS). We follow the description of those data with a summary of results that examines the causes of species decline in the USA, and suggestions concerning the implications of this work for further studies of species decline in green criminology.

Background: Green criminology, capitalism and species decline

Green criminology has focused much of its analysis of threatened and endangered species on case studies of specific animals (e.g. Hagstedt and Korsell, 2012; Wyatt, 2009, 2011) rather than empirical approaches (Clarke and Rolf, 2013) to address species decline and endangerment issues. To date, both case studies and empirical studies have been compiled on a case-by-case or species-by-species basis. Extant research of this nature has provided important insights into certain aspects of anthropocentric harms that impact single species. These studies tend to focus almost entirely on the negative effects of poaching and hunting (for an exception, see Clarke and Rolf, 2013) and tend to be species specific, impeding a broader analysis of the commonalities among crimes/harms against non-human species.

The vast majority of species that are threatened or endangered, however, find themselves in such a state of endangerment due to the more general effect of human development on ecosystems (Czech, 2000;

Czech et al., 2000; Naidoo and Adamowicz, 2001) rather than as a result of more localized human behaviours, such as poaching and hunting. The widespread nature of non-human species endangerment indicates that broader structural processes have more salient and persistent effects across species and nations. The vast majority of species that are recognized as threatened or endangered legally are those that are impacted by various forms of human development rather than through poaching, hunting or animal trade. Human development has widespread impacts on species by destroying ecosystems in ways that sometimes eliminates these ecosystems and non-human species locally, and on a larger scale impeding ecosystem functionality and habitat structures through processes such as ecosystem segmentation that have negative impacts on non-human species viability. In a structural view the impacts of various forms of development are generally much more significant for species than activities such as poaching, hunting and animal trade. For example, Woodroffe (2000: 168) demonstrates a strong correlation between economic development and carnivore extinction, noting that 'for most of the species, local extinctions are associated with growing human populations'. In a significant assessment of the state of global extinction research and rates, Stork (2010) does not mention poaching once. Nevertheless, poaching, illegal hunting and illegal trade in animals have tended to attract the attention of criminologists because these activities are clearly defined as illegal and thus are identified as crimes in some form of national or international law (for alternatives, see Sollund, 2011, 2013). In taking that approach and focusing on behaviours that are defined as illegal in law, green criminologists have been able to show that the harms that they are exploring fit within the scope of more traditional forms of criminological analysis since they involve violations of law (e.g. Clifford, 1998; Franz, 2011; Greife and Stretesky, 2013; Pires and Clark, 2012). At the same time, this focus on violations of specific laws that define non-human species harms as crimes tends to overlook important structural forces that impact non-human species viability.

For specific species, the effect of poaching can be dramatic, especially when coupled with the long-term impacts of development, so we are not suggesting that green criminologists abandon their analyses of crimes such as poaching (i.e. abduction and killing; see Sollund, 2011). However, greater attention must be paid to the ways in which illegal behaviours such as poaching intersect with structural dynamics to produce large-scale non-human species harms. One example of the combination of these processes (e.g. poaching and structural dynamics) can be seen in studies of the tiger population, which is significantly impacted by poaching and more general structural conditions. The World Wide Fund for Nature estimates that the world's tiger population now occurs in only 7% of its historic range and comprises approximately 3,200 individuals, significantly less than the estimate of 100,000 for tigers in 1900 (see also Chundawat et al., 2013; Dinerstein et al., 2007). It is estimated that in recent years the number of tigers poached from the wild is somewhere between 200 and 250 individuals per year. For a population of 3,200 individuals, poaching certainly produces a significant impact on species viability and, if these estimates are correct, it leads to a loss of 6% or more of the tiger population annually. Current estimates suggest that if tiger-poaching patterns remain unchanged, the tiger is likely to 'disappear from many more places, or dwindle to the point of ecological extinction' (Dinerstein et al., 2007: 513). We must recognize that factors other than poaching also contribute to declining tiger populations, and may significantly influence tiger survival and the probability of extinction. These factors include human development effects associated with the destruction and fragmentation of tiger habitat through deforestation, the effect of habitat destruction on tiger reproduction rates, the loss of tigers through human–tiger conflict, and the loss of tiger prey species due to habitat destruction (Smith, Ahern and McDougal, 1998; Karanath and Stith, 1999; Kenney et al., 1995; Linkie et al., 2006).

For the vast majority of endangered non-human species, it is human development and the encroachment on natural ecosystem space rather than poaching, hunting or animal trade that leads to their listing as a threatened or endangered species. For example, when deforestation occurs, the impact is felt across a range of species that make their homes in forest habitats. While we may single out particular species for attention (e.g. tigers, elephants, pandas or whales), there are large numbers of species to which we pay little attention (e.g. fungi, insects and velvet worms, see Feuerer and Hawskworth, 2007; Hawksworth, 2001) that are also impacted by forms of human development that imperil wild areas. In particular, in our view the driving force behind widespread non-human species habitat destruction is the nature of the global capitalist world system and the normal operation of the capitalist treadmill of production (see Schnaiberg, 1980; Gould et al., 2008), which significantly influences ecosystem destruction across nations of the world in ways that are consistent with the structural organization of global capitalism (see also Stretesky et al., 2013).

In the contemporary era, non-human species are becoming extinct at extraordinary rates. In the next section we briefly make the case that non-human species harm is widespread and often overlooked.

Anthropocene extinction and capitalism

We live in a rapidly changing world – one where human development continuously creates and spreads ecological destruction. One indicator of the extent and widespread nature of such harm is the human ecological footprint, which is now approximately 1.5, meaning that humans are consuming nature 1.5 times as fast as nature can reproduce itself (Wackernagel and William, 2012). That rate of consumption also means that humans are rapidly eroding the ability of the ecosystem to support not only humans but non-human species as well, placing them in conditions that promote accelerated rates of extinction.

One of the responses to species endangerment is the creation of laws that identify threatened and endangered species and single them out for protection. Such statutes began to emerge broadly in the 1970s. Species are officially recognized as endangered to facilitate the implementation of additional protective policies that are designed to prevent extinction. Nevertheless, contemporary extinction rates for animal and plant species are quite high and exceed the background rates of extinction – that is, the natural rate of extinction under conditions when humans did not exist and could not impact ecosystems and species viability (Lomolino et al., 2001). Of particular concern is the fact that extinction rates since the industrial era are so high that researchers have identified this period as the sixth wave of extinction and have named it the Anthropocene to identify the fact that extinction in this era is driven by human influences (Barnosky et al., 2011; Steffen et al., 2007, 2011; Zalasiewicz et al., 2010). Evidence supporting the Anthropocene extinction has identified specific influential factors that are associated with human development and ecological impacts on species as responsible for the current extinction rates. These factors include deforestation, climate change (Parmesan and Yohe, 2003; Thomas et al., 2003; Thuiller et al., 2005), the expansion of the ecological footprint and accelerating consumption (Vackar, 2012).

From the preceding it can be argued that the widespread effect of human development on species often occurs through routes that do not necessarily involve illegal activities such as poaching or illegal animal trade. Rather, species endangerment and extinction are largely the result of routine human activities that are related to economic development and expansion that we suggest is part of the expansionary tendencies of capitalism. Theoretically, human development and the consumption of nature through ecological withdrawals, ecological additions and ecosystem space conversion create the structural conditions through which ecosystems and consequently species are destroyed. In this sense, then, we can say that the majority of species that become extinct or which are listed as threatened and endangered are a consequence of 'normal' patterns of human economic development. In the modern era, those 'normal' patterns of development are associated with the constant expansion of the economy, and thus can be interpreted as a consequence of the expansion of capitalism and the capitalist treadmill of production (Stretesky et al., 2013).

Dating back hundreds of years, capitalism emerged as the dominant economic form in the world market, exerting extensive negative ecological pressure and destruction, and limiting ecosystem functionality (Burns et al., 1994; Hornborg, 1998; Jorgenson, 2004). Thus to understand ecological consequences such as the widespread decline of various species that have been identified as threatened, endangered or extinct in the Anthropocene, it is necessary to refer to the ways in which capitalism produces ecological pressures that facilitate the decline of species. Here we begin with the widely recognized observation that capitalism's primary goal is the production of continually expanding profit. To meet this goal, the capitalist system of production must constantly increase production, and hence must also constantly increase its extraction and consumption of raw materials. The result of constantly expanding production and consumption is the acceleration and expansion of ecological destruction and disorganization (Stretesky et al., 2013). These negative ecological impacts of capitalism are widely detailed and explained in the ecological literature (Burkett and Foster, 2006; Clark and York, 2005; Clausen and Clark, 2005; Foster, 1992, 1994, 1997, 1999, 2000; Foster and Clark, 2004; Jorgenson, 2004; Foster et al., 2010; Jorgenson and Clark, 2011; Schnaiberg, 1980). Moreover, these assertions about capitalism and ecological destruction have extensive empirical support. Empirical examinations of the capitalism–species endangerment connection support a link between capitalism and species destruction across nations (Brewer et al., 2012; Clausen and York, 2008; Hoffmann, 2004; McKinney et al., ,2010; Shandra et al., 2008), and with processes that are related to species endangerment, such as deforestation (Jorgenson, 2006).

Here we hypothesize but do not test the assertion that the widespread nature of species decline over time and across nations must be the result of influences or factors that occur persistently across nations and over time. They must also be factors that apply across a diverse range of non-human animal and plant species to effectively explain the widespread nature of species decline across societies, time and species.

These influences are beyond the scope of activities such as poaching, hunting and animal trade. For example, a number of insects worldwide are listed as endangered. An example of several US insect species that are listed as endangered or critically endangered by the IUCN or the USFWS is as follows: American Burying Beetle; Avalon Hairstreak; Blind Cave Beetle; Carson's Wandering Skipper; Columbia Clubtail; Delhi Sands Flower-Loving Fly; Desert Everglade Sprite; *Dorymyrmex insanus*; Everglades Sprite; Franklin's Bumble Bee; Helotes Mold Beetle; Keys Scaly Cricket; *Manica parasitica*. Besides the fact that these creatures may be collected for scientific purposes by those with appropriate licences or by a curious youngster, they have no commercial value and are not trafficked, poached or hunted, yet they are still threatened with extinction. These geographically diverse species are endangered by forms of human development that encroach upon and destroy natural spaces. That pattern of human encroachment is, we suggest, related to the continuous expansion of the capitalist treadmill of production. That is, while some non-human species certainly became extinct prior to capitalism, they are doing so at a greater rate following its development – and clearly are doing so independently of population growth. This occurs in two ways. First, researchers who study species extinction demonstrate that the expansion of logging contributes greatly to these non-human rates of extinction and decline. For example, in the journal *Nature*, Pimm and Raven (2000) point out that damage to timber hotspots is the single greatest threat to non-human loss, and they emphasize that it is a relatively recent phenomenon. That is, as timber was needed as an energy source for the development of capitalism, a 'relaxation' period occurred. Relaxation needs to be taken seriously as a source of non-human extinction and should not be discounted as similar to the causes of extinction that occurred prior to the development of capitalism. Relaxation suggests that the 'the original number of species in the fragmented area eventually relaxes to a new, lower number' (Brooks et al., 1999: 1140). While there is debate about the timeframe associated with the emergence of a relaxation period, the dates of timber withdrawals suggests that the significant expansion of timber-related energy that were needed to drive capitalist production are the cause. This is especially documented during the period of 'advanced industrialization' that occurred worldwide between 1920 and 1960 (Tillman, 2012). More importantly, current background levels as a result of this capitalist expansion are driving extinction rates to the tune of at 'least 1000 times higher than the background rate' without this form of human economic activity (Brooks et al., 1999: 1150).

Second, it is also increasingly important to note that as populations expand, they have destroyed species to provide for continual economic growth. We need to be clear that it is not population growth directly that is the cause of this destruction. That is, population expansion and growth in the current period are organized to be compatible with the expansion of the treadmill of production. Thus we take a similar position as Tabb and Sawers (1984: 4), who point out that living spaces are 'merely a reflection of the larger economic and social fabric, termed the mode of production'. As a result, the organization of various living spaces that may encroach on natural habitats is likely to be environmentally destructive in the current period of capitalism in different ways than those developments that occurred prior to the emergence of capitalism where the human ecological footprint was at an all time low. This issue is made clear by biologists who study the history of extinction rates. It is not population that matters but the ecological impact of that population under the current form of production. Thus currently we see extremely high ecological footprints in those nations that are most central to the production process (see Jorgenson, 2003). For example, Barnoski et al. (2011: 57) point out that this problem is directly linked to what some biologists have called the sixth mass extinction:

> there are clear indications that losing species now in the 'critically endangered' category would propel the world to a state of mass extinction that has previously been seen only five times in about 540 million years. Additional losses of species in the 'endangered' and 'vulnerable' categories could accomplish the sixth mass extinction in just a few centuries. It may be of particular concern that this extinction trajectory would play out under conditions that resemble the 'perfect storm' that coincided with past mass extinctions: multiple, atypical high-intensity ecological stressors, including rapid, unusual climate change and highly elevated atmospheric CO_2.

In short, as Barnowski et al. point out, it is the organization of society that is the threat. To further illustrate the widespread nature of the species-endangerment problem, below we review data on the count of endangered species globally and within the USA. While we do not test our proposition that these widespread patterns of endangerment are a product of capitalism with these data, future research should be explored that addresses this issue. For example, statistical models of population extinction could be used to test the difference in extinction coefficients that are associated with variations in modes of production. The ubiquitous nature of capitalism, however, makes such hypothesis tests difficult. Thus further refinement of an applicable hypothesis concerning variations in core aspects of capitalism are needed before such a hypothesis can be tested with these data.

Global count of endangered species: IUCN Red List

Threatened and endangered species are found in all nations of the world. Efforts to count and regulate them at the international and national levels exist and provide us with some evidence of the widespread nature of the problem.

In 2012 the IUCN updated its Red List, which is an assessment of the health of species derived from its global survey. Part of the data keep track of species extinction and the status of endangered species. IUCN estimates that there are currently 1,729,693 species in various categories that exist throughout the world.[3] Of those species the status of only 71,576 (4.1%) has been adequately assessed by the IUCN. Of the assessed species, 21,286 (29.7%) were identified as threatened in 2013. Of the threatened species, 6,451(30.3% of threatened species and 9.1% of all assessed species) are listed as endangered, while 4,286 species (20.1% of threatened species and 6% of all assessed species) are listed as critically endangered.[4] In addition, 61 species are listed as extinct in the wild while an additional 799 species are listed as extinct. Thus of the 71,576 species surveyed in the Red List, about 15.1 % are identified as endangered/critically endangered while 1.2% of all identified and studied species are listed as extinct.

Table 7.1 summarizes the global IUCN data reports (IUCN, 2013). It shows a collapsed distribution of species in four categories:

Table 7.1   IUCN species data summary

	Estimated	Assessed	Threatened 1996	Threatened 2013
Vertebrates	65,146	37,356	3,314	7,390
Invertebrates	1,305,250	15,911	1,891	3,822
Plants	306,674	18,291	5,328	10,065
Fungi/protists	51,623	18	–	9
TOTAL	1,729,693	71,576	10,533	21,286

Notes: Vertebrates include mammals, birds, reptiles, amphibians and fish. Invertebrates include insects, molluscs, crustaceans, corals, arachnids, velvet worms and horseshoe crabs. Plants include mosses, ferns, gymnosperms, flowering plants, green algae and red algae. Fungi/protists include lichens, mushrooms and brown algae.

vertebrates, invertebrates, plants and fungi/protists. Of the species in each category, the status of 57.3% of vertebrates; 1.2% of invertebrates; 6.0% of plants and 0.04% of fungi/protists has been assessed (for a discussion, see note 2). For vertebrates, 19.8% of species are classified as threatened; for invertebrates, 24.0%; for plants, 55.0%; and for fungi/protists, 50%.

Research on the status of various species indicates that many are in a state of decline. Stuart et al. (2004) note that scientists have been examining declines in amphibian populations since the 1970s. Employing IUCN data, they examined the status of the then identified 5,743 (2004) amphibians (or 81.5% of the number of amphibians that the IUCN currently identifies). One of the limitations of this study was that insufficient data were available for 22.5% of amphibians. Stuart et al. found that amphibians were more widely threatened than mammals or birds. Since 1980 sufficient data has accumulated to indicate that nine amphibians have become extinct, while other data indicate that up to 113 additional amphibians can no longer be located in the wild (Stuart et al., 2004). One of the major causes of amphibian decline is reduced habitat. However, many of the declines are classified as 'enigmatic', meaning that the cause is not certain. Of the 435 species identified as facing a strong threat of extinction in 2004 compared with 1980, 50 (11.5%) faced population declines due to 'overexploitation' or extraction from natural environments; 183 (42.1%) due to habitat destruction; and 207 (47.6%) due to enigmatic forces, which have largely been associated with diseases and climate change. As these data indicate, the amphibian trade, poaching and illegal trafficking (overexploitation) play a relatively minor role in the decline of amphibians worldwide. Similar conclusions have been reached with respect to global bird (Sekercioglu et al., 2008) and reptile populations (Gibbon et al., 2000). This is an important point to consider, since much of the green criminological research on species declines have focused on population declines that are related to overexploitation as opposed to those associated with habitat destruction or enigmatic forces, such as climate change (for an alternative discussion, see Sollund, 2012b). That focus has important implications for the kinds of policy response that green criminologists suggest for controlling biodiversity loss and species declines. While species decline and biodiversity loss due to overexploitation are important and should not be overlooked, the major forces behind both are habitat loss and climate change, and these require different types of control policy from that are often suggested by green criminologists.

In the section that follows we examine the distribution of endangered species across the USA. The data examined are extracted from the USFWS. Data on extinct species that were once located in the USA are also examined. These data were collected from USFWS information and trace extinct species since 1860, including some species that were identified as endangered or critically endangered by the IUCN.[5]

The distribution of endangered species across the USA

Above we have reviewed IUCN data on endangered species which focus on the global nature of that problem. We also wish to draw attention to the problem of endangered species on more local levels. For this purpose we examined data on threatened species across the states of the USA.

The USFWS publishes a list of threatened species in the USA following the specifications of threatened species as identified in the applicable US laws (50 CFR 17.11(h) and/or 50 CFR 17.12(h)). Based on those laws there are currently 645 animal species and 872 plants species that are listed as threatened or endangered in the USA (USFWS, 2013). The number of threatened and endangered species of animals and plants by state is shown in Table 7.2.

Table 7.2   Endangered and threatened species count by state, USFWS

	Animal 1	Animal 2	Plant 1	Plant 2	Total	ESEF*
Alabama	103	3	18	1	125	8.24
Alaska	16	1	1	0	18	1.19
Arizona	37	1	20	1	59	3.89
Arkansas	24	5	5	0	34	2.24
California	124	2	180	5	311	20.50
Colorado	16	0	16	0	32	2.11
Connecticut	14	0	2	0	16	1.06
Delaware	10	0	5	0	15	0.99
Florida	64	3	57	0	114	7.52
Georgia	38	2	22	2	64	4.22
Hawaii	65	2	362	6	435	28.68
Idaho	9	0	3	0	12	0.79
Illinois	19	0	9	0	28	1.85
Indiana	17	0	4	0	21	1.38
Iowa	9	0	5	0	14	0.92
Kansas	11	2	2	0	15	0.99
Kentucky	35	1	8	0	44	2.90
Louisiana	18	4	3	0	25	1.65

	Animal 1	Animal 2	Plant 1	Plant 2	Total	ESEF
Maine	9	2	3	0	14	0.92
Maryland	16	0	6	0	22	1.45
Massachusetts	17	1	3	0	21	1.38
Michigan	13	0	4	0	21	1.38
Minnesota	10	0	4	0	14	0.92
Mississippi	33	8	4	0	45	2.97
Missouri	25	2	10	0	37	2.44
Montana	9	0	3	0	12	0.79
Nebraska	8	3	4	0	15	0.99
Nevada	27	3	9	0	39	2.57
New Hampshire	5	3	3	0	11	0.73
New Jersey	13	1	6	0	20	1.32
New Mexico	31	3	13	0	47	3.10
New York	17	1	7	1	26	1.71
North Carolina	29	6	27	0	62	4.09
North Dakota	6	0	1	0	7	0.46
Ohio	17	0	5	1	23	1.52
Oklahoma	18	1	1	1	21	1.38
Oregon	36	7	16	2	61	4.02
Pennsylvania	11	0	2	0	13	0.86
Rhode Island	10	1	2	0	13	0.86
South Carolina	15	5	19	2	41	2.70
South Dakota	9	1	1	0	11	0.73
Tennessee	71	1	18	1	91	6.00
Texas	60	3	30	1	94	6.20
Utah	16	1	25	0	42	2.77
Vermont	2	0	2	0	4	0.26
Virginia	51	5	15	3	74	4.88
Washington	30	3	9	0	42	2.77
West Virginia	14	2	6	0	22	1.45
Wisconsin	9	1	7	0	17	1.12
Wyoming	6	0	4	0	10	0.66
TOTAL	1262	90	995	27	2374	156.52
Mean	25.24	1.8	19.9	0.54	47.48	3.13
Range (low)	2	0	1	0	4	0.26
Range (high)	124	8	362	6	435	28.68

Notes: Animals include mammals, fish, amphibians, reptiles, birds and insects.
Animal 1 = endangered and threatened animals specifically recognized in each state.
Animal 2 = endangered and threatened animals that occur in state but are not listed as state species.
Plant 1 = endangered and threatened plants specifically recognized in each state.
Plant 2 = endangered and threatened species that occur in state but are not listed as state species.
ESEF = endangered species enrichment factor. This measure was created by dividing the total number of threatened and endangered species in a state by the total number of endangered and threatened species in the USA (N = 1,517).
Source: USFWS (2013).

To succinctly summarize the USFWS data, we collapsed them into two general categories: animals and plants. The table shows two categories of endangered species for animals and plants. The 'Animal 1/Plant 1' columns shows the number of threatened species that are officially recognized in each state. The 'Animal 2/Plant 2' columns shows the number of species that, while not officially recognized in each state, also appear in those states. In our calculations the sum of Animal 1, Animal 2, Plant 1 and Plant 2 was used to identify the total number of animal and plant species that are at risk in each state and across states.

Following US law the USFWS recognizes the existence of 1,517 endangered plant and animal species. The total number of endangered species listed in Table 7.2 (the sum of the 'Total column') exceeds that figure because a species can be listed in more than one state and species listed in multiple states are thus counted more than once. Across states the total number of endangered plant and animal species is 2,374 (1,352 animals and 1,022 plants) rather than 1,517.

The data in Table 7.2 can be used to describe some characteristics of the distribution of endangered species across the USA. The range for endangered animal species, for instance, is between 2 (Vermont) and 126 (California), with a mean of 27.04 endangered species per state. For plant species the range is between 1 (Alaska) and 368 (Hawaii), with a mean of 20.44 endangered species per state. The mean number of endangered animal and plants species per state is 47.48. That per state average is significantly impacted by California and Hawaii. For animal species, 19.5% of 645 endangered animal species occur in California; 42% of the 872 endangered plants species occur in Hawaii. Thus, when omitting California for animal species, the mean number of endangered species per state declines to 25, or by about 7.2%. Omitting plant species in Hawaii has a much greater effect, and the mean number of endangered plant species across states declines from 20.44 to 13.3. For both plant and animal species, omitting both California and Hawaii decreases the mean number of endangered species per state from 47.48 to 33.9, or by 28.6%.

The distribution of endangered species is highly skewed and uneven across states. The majority of states (N = 29 or 58%) have a distribution of endangered species of between 4 and 27. Within that small range of endangered species and states, the distribution approximates a normal curve. For the remaining 21 states, however, the distribution of endangered species ranges from 28 to 435, and with the exception of small spikes around 38–47 (N = 7), 58–62 (N = 3) and above 100 (N = 3), the number of endangered species is widely distributed.

In the final column in Table 7.2 we present what we call the endangered species enrichment factor (ESEF). This is the percentage of all recognized endangered species (N = 1,517) found within each state. At the low end, only 0.26% of endangered species are located in Vermont. At the upper end, 28.68% (plant) and 20.5% (animal) of endangered species are located in Hawaii and California, respectively. The ESEF therefore indicates that both Hawaii and California appear to be locations of significant species diversity and endangerment. With respect to policy and efficiency, one could make the argument that efforts to control human ecological destruction in Hawaii and California has more 'bang for the buck' with respect to the probability of aiding in protecting endangered species. Such an approach, however, would overlook the fact that in other kinds of ecosystem, important ecosystem species are in danger of becoming extinct and threatening the functioning of the ecosystems in those locations as well.

These data have their limitations. As Wilcove and Master (2005) note, the vast majority of species in the USA have not been 'well studied', meaning that there is insufficient data to efficiently judge the extent to which species are threatened and endangered in the USA. Reviewing the available data, Wilcove and Master estimate that the number of species listed in the Endangered Species Act (ESA) is well below the level that is needed to protect species biodiversity in the USA. They suggest that the ESA list of species needs to increase by a factor of at least ten to efficiently protect species and to adequately represent the extent of species endangerment in the USA.

Having reviewed the extensive variation in species endangerment for species at the global level and across the USA, in the next section we address the relevance of this exploratory study to future green criminological research.

Discussion and conclusion

Based upon the theoretical discussion and the data presented above, we suggest that green criminologists should devote additional attention to the use of empirical data that address the relationship between anthropogenic sources of ecological disorganization and species endangerment, and to structural processes that influence harms against non-human species across and within nations. To date the majority of green criminological studies on biodiversity loss have used case studies and qualitative approaches which, while informative for any particular individual species, do not provide an adequate understanding of broader structural factors that influence species and biodiversity loss across the globe. Like several other green criminologists, we argue that these structural factors are largely responsible for environmental harm (Ruggiero, 2013; Ruggiero and South, 2010; Walters, 2010). In our view those structural factors are specifically related to the organization of capitalism, its inherent drive to constantly expand production and, as a result, its continual need to escalate the extraction of natural resources for the creation of commodities.

The data examined above indicate the need for further research that is focused on identifying the factors that may be associated with the distribution of endangered species across the USA. Persistent and widespread effects on non-human loss are produced by habitat loss and climate change. The effects of poaching are much more limited. Even for African elephants, a species that suffers from extensive poaching, this activity leads to an estimated loss of 7.4% of elephants (Lawson and Vines, 2014). Clearly one of the important factors in species loss is their natural distribution, and species are unlikely to be lost at high rates under certain distribution parameters. The natural distribution of species, however, does not tell us why they become endangered at any specific location. Species endangerment is a consequence of the intersection of species distribution with factors that promote endangerment. The latter include those identified above: enigmatic effects such as climate change which may be difficult to isolate at the state level; overexploitation of species and variability in overexploitation in different regions; habitat destruction; and the general effects of the expansion of the capitalist treadmill of production and its impacts on ecological disorganization through ecological withdrawals and additions (see below). We limit our discussion to examples of empirical research that have been conducted in the USA to focus attention on the USFW service data reviewed above.

Prior research on the distribution of endangered species across the USA has found that there are hotspots of biodiversity loss and that these ought to be addressed when devising policies to control such loss (Dobson et al., 1997; see also Czech and Krausman, 1997; for additional hotspot analysis on biodiversity, see Orme et al., 2005). Non-human species loss hotspots are those that are associated with human development and encroachment on natural environments, which we see as a result of the constantly expanding nature of the capitalism treadmill of production.

Theoretically, loss hotspots are important with respect to policy issues. They are useful to the extent that they can be targeted to improve the efficiency of species-protection efforts. At the same time we must keep in mind the possibility that a focus on loss hotspots to the exclusion of larger species protection policies will be insufficient to broadly protect species from economically generated harms that cause biodiversity and species loss that stem from economic development promoted by the treadmill of production. Related studies should also be considered, especially from a localized hotspot perspective. Ando et al. (1998), for example, found that land value plays a significant role in being able to institute efficient conservation policies that protect endangered species, perhaps indicating that endangered species are better protected in states where a significant volume of low economic value land is present and available for use in conservation efforts and where, therefore, large tracts of land can be purchased that will protect numerous species simultaneously.[6] In those locations the differential development of the treadmill of production is likely to impact land values. As Marxist ecologists such as Foster et al. (2010) note, the long history of the metabolic rift between urban and rural areas establishes an unequal exchange of metabolic materials from rural to urban areas, resulting in forms of ecological destruction in rural areas (e.g. dams, and mountaintop removal in the case of coal extraction). That form of ecological destruction can reduce land values, as rural land adversely impacted by the treadmill of production practices loses its natural productivity. Illustrating the utility of this approach are studies that examine species loss from the opposing direction as well – that is, studies that draw attention to the relationship between habitat loss and species loss. Wilcove et al. (1998) examined the effects of several variables that are relevant to the modification of ecosystems that included habitat destruction but also measures of over-exploitation of natural resource systems, the occurrence of alien species, and disease as threats to endangered species survival. Consistent with the above observations, they found that habitat loss had the strongest effect on species endangerment. Such results indicate the need for green criminologists to pay additional attention to concerns such as habitat loss as a major driver behind species endangerment and extinction.

Other studies point to the effects of economic development on species diversity. In our view, such studies are important because the capitalist treadmill of production drives economic development to continually expand in the pursuit of profit regardless of the ecosystem consequences. Importantly, as Czech (2000) noted, ecologists have paid insufficient attention to the effect of economic development and growth on species and wildlife conservation, and some, influenced by inaccurate depictions of the effect of indigenous peoples on species loss, follow reasoning which suggests that it is the poor people of the world and population growth in those regions that drive species extinction. In contrast, a structural economic view calls attention to the fact that the traditional behaviours of indigenous peoples only become ecologically destructive once capitalism has expanded into developed regions and claims a significant portion of the ecosystem for production, especially through the extraction of raw materials and imposed monoagricultural methods of production that are consistent with the efficiency requirements of capitalism. Following up on the economic development hypothesis, Czech et al. (2000) found that economic growth had an marked impact on species endangerment, which in our view indicates the potential importance of considering the impact of the treadmill of production on species endangerment. Following Czech's (2000) critique, one can suggest that, like some ecologists, green criminologists have not paid sufficient attention to the effect of economic development on species endangerment and extinction. The more general ecological literature, however, indicates that the greater concerns with respect to species health and vitality are related to economic development and habitat loss – issues recently raised, for example, in discussions of the treadmill of crime (Stretesky et al., 2013).

Studies also indicate the importance of considering other indicators of development on species endangerment. Though not restricted to studies performed in the USA, Luck (2007) undertook a meta-analysis of studies that examine the relationship between human population density and species diversity, and found that the evidence for a link between human population density and biodiversity loss was weak. In a cross-state model, Brown and Laband (2006) assessed the relationship between human activities that impact ecosystems (specifically mean household population density, an indicator of roadways, and the intensity of nighttime lighting) and species endangerment. Controlling for population density, they found that every 1% increase in human activity leads to an increase in species endangerment of 0.25% across states. In their study of variations across states for federally listed endangered mammals, Kirkland and Ostfield (1999) found significant effects on habitat diversity, wetland loss, percentage of forest cover and area reserved for state parks.

At the cross-national level, Naidoo and Adamowicz (2001) tested the effects of per-capita gross national product on the number of threatened species across nations. Their study was devised as a test of the environmental Kuznets curve (EKC) argument that over time or across nations, economic development has an inverted 'U' shape relationship to ecological destruction – that is, that after per-capita income reaches a saturation point, additional economic development causes a reduction in ecological destruction. That argument suggests that, initially, economic development has a negative impact on ecological destruction, so that as economic development expands, ecological destruction increases which, in Naidoo and Adamowicz's view, would increase species endangerment. However, the EKC argument also suggests that once a certain point in economic development is reached, the association between economic development and ecological destruction reverses, and further economic development becomes a protective factor that diminishes ecological destruction and, therefore, reduces species endangerment. Naidoo and Adamowicz raise the question as to whether the traditional EKC argument applies to endangered species. Their findings suggest that for the majority of taxonomic species (excluding birds) the observed relationships rejected EKC assumptions, and that economic development increases species endangerment instead of protecting species.

In closing, we return our attention to the issue of structural explanations of non-human species harm and decline. As we have illustrated, endangered species are widely distributed and they include a range of non-human species. Across these species we argue that the main cause of endangerment to non-human species is capitalism. Thus explaining and understanding the scale of species endangerment requires the adoption of structural views of non-human species endangerment and extinction. Specifically, we take an alternative perspective on species extinction and argue that extinction patterns are not caused by population expansion, poaching and/or wildlife trade. For example, as noted above, even for species in which poaching plays an important role in the final stages of their decline (e.g. elephants), this activity accounts for only a small percentage of wildlife loss (i.e. in the study cited above, 7.4% of elephants are lost to poaching). Given the available data such as those on the percentage of elephant losses due to poaching, the assumption that it drives species extinction places undue emphasis on the activity as a cause of extinction and, we would also suggest, draws attention to the behaviour of indigenous peoples as the cause of such a problem, while simultaneously diverting attention from the structural economic origins of species extinction.

In lieu of arguments that 'blame the poor' and developing nations for ecological problems such as species extinction, we opt instead for a structural explanation. This has an advantage with respect to species endangerment and extinction related to explaining the pattern of species extinction. For example, the data from the USA on species endangerment by state are useful to illustrate our point. Within the USA there is no division between states that is similar to the division between nations (developed vs. developing). That is, no state is so economically disadvantaged that its level of income shrinks to levels that are found in developing nations. For example, Santos-Paulino

(2012) has pointed out that 80% of the population in the United Nations sample of developing nations exist on an income of less than US$ 2 per day. Returning to the USA, there is no indication that under-development impacts endangerment or extinction. We also suggest that such a hypothesis would not be applicable within the context of any developed nations. Thus, framing species extinction simply as a problem caused by people in underdeveloped nations misses the larger point concerning the distribution of species endangerment and extinction across other areas of the world.

As we have argued, it is our contention that levels of endangerment and extinction are connected to the expansion of capitalism in its current form. For example, the Amazon basin has a high rate of threatened and endangered species (van Solinge, 2008). Most of the countries in that region are classified as developing and some as underdeveloped economically. However, the simple fact that the high level of species endangerment and extinction co-occur in nations that are economically underdeveloped does not, in turn, mean that species extinction and endangerment are driven by economic underdevelopment or the specific behaviours of people living in those nations. The correlation is spurious, as many residents of that basin would suggest. On this point our structural economic argument draws attention to the fact that economic underdevelopment in the Amazon is linked to its exploitation and the development of the global treadmill of production. It is widely recognized that capital penetration into the Amazon region plays a very significant role in ecological destruction there, and that includes treadmill of production processes related to the extraction of timber and its effects on deforestation, as well as treadmill of production agricultural practices that convert forest lands to agricultural use, including cattle farming, to produce food products for those in the developed nations (Boekhout van Solinge, 2008, 2010).

We recognize that there are many processes that contribute to species extinction and threats. Those factors that we have not explored are wars, pesticide use, and abductions and killings by humans. Moreover, we hold that it is entirely possible to enumerate each of these contributing factors – that is, relative weights can be assigned to species endangerment. The point we are making, while somewhat controversial, is that the majority of the factors that contribute to endangerment and extinction are so mathematically small that they cannot be compared to the impact of the treadmill of production and capitalist economic development. Part of the point in drawing attention to this connection is to suggest that criminologists should pay closer attention to how economic structures, such as the treadmill of production, are the chief drivers behind the process of species decline in the contemporary era. As we have noted, the scientific literature supports our theoretical assertions. That is, scientists are now beginning to talk about a new era of species extinction which ranks among the most significant in biosphere history. The difference is that this time the loss in biodiversity is a result of the way in which human economics interact with the ecology. This relationship has produced the term 'Anthropocene' to indicate that species extinction is driven by human behaviour. Above all, we argue that the scientific literature recognizes that the single most important human effect on rates of extinction and endangerment is economic development. We believe that if the discipline of green criminology does not take science seriously, that a criminological literature will emerge that is inconsistent with empirical scientific evidence on this point. That is, an emphasis within criminology will develop that focuses on the least important variables as the drivers of destruction.

We have argued that one theoretical perspective that can be employed to highlight the connection between ecological concerns, such as species extinction/threats and economic development, is to pay greater attention to the past 34 years of research on the intersection of economic development and ecological destruction found in the ecological economics literature. From among the options in that literature we have focused our attention on the specific connection between economic development and ecological destruction presented in the treadmill of production literature and the ecological Marxist literature. As noted above, both contain substantial empirical support for the proposition that economic development and ecological destruction are linked. In the present analysis we have extended those economic arguments to threats to non-human species as one form of ecological destruction, and have illustrated points where the scientific literature would also support such a contention.

One of our aims is therefore to draw attention to the fact that outside criminology there are numerous empirical studies that are directly relevant to the interaction between economic development and species threats/extinction. They call attention to the persistence of these threats to species globally. We see these studies as indicating a need for a structural analysis of the threat to species that is situated in ecological Marxism (e.g. Burkett and Foster, 2006; Foster, 2000) and the treadmill of production as developed by Alan Schnaiberg (1980). Absent efforts to understand the structural nature of non-human species decline, green criminologists will be confined to a case-study approach, whether empirical or qualitative, and will fail to come to grips with the broader economic forces that drive species harm.

Notes

1. Where we use the term 'non-human' we are generally referring to non-human animal species. Non-human non-animal species are also the victims of systemic ecological harm but are excluded here to constrain the size of the current project.
2. The combination of the IUCN Red List species and the USFWS lists include the following wildlife as animals and plants: (animals) mammals, birds, reptiles, amphibians, insects, molluscs, crustaceans, corals, arachnids and horseshoe crabs; (plants) mosses, conifers, sponges, ferns and allies, gymnosperms, flowering plants, trees, green algae, red algae, lichens, mushrooms and brown algae.
3. Estimates of the number of known species are controversial. Consider, for instance, consider that concern for the number of fungi species. The Red List estimates 51,623 as of 2013. The literature on fungi species, however, accepts an upper end estimate of approximately 1.5 million (Hawksworth 2001), or 29 times the number of fungi species that the IUCN recognizes. In contrast, Frohlich and Hyde (1999) suggest that the number may be as high as 9.9 million. In addition, the literature suggests that the minimum number of fungi species is at least 74,000 (Hawksworth 2001), a figure that is still 43% higher than the Red List estimate. Thus the Red List appears to underestimate the number of fungi species. With respect to lichens, Feuerer and Hawksworth (2007) employ a checklist measure which suggests that there are 18,882 lichens compared with the Red List estimate of 17,000.
4. For specific details about the definition of terms that are used for Red List classifications, see ICUN (2013). Definitions of 'endangered' and 'critically endangered' involve complex measures. Here we include a summary for critically endangered and endangered species (however, this does not provide a sufficient indicator of the complexity of the measurements involved):

	Critically Endangered	Endangered
Population size measure		
Population reduction observed, estimated or suspected	>90%	>70%
Geographic range		
(A) Extent	<100 km^2	<5,000 km^2
(B) Area of occupancy	=1	<5
Number of mature individuals	<250	<2,500
(C) Projected continuing decline	25% in 3 years	20% in 5 years
(D) Number of mature individuals		
In subpopulations, or	<50	<250
(E) % mature individuals in one Subpopulation	90%–100%	95%–100%
Very small/restricted population	<50	<250
Probability of extinction in the wild	>50% in 10 years	>20% in 20 years

5. The extinct, and possibly extinct, species for the USA (N = 142) and the year of extinction since 1860 are as follows, as taken from the IUCN Red List, USFWS data and Fuller's *Extinct Birds* (2000; NY: Oxford University Press): (1) Acorn Pearly Mussel, unknown; (2) Acorn Ramshell, unknown; (3) Agate Rocksnail, unknown; (4) Alabama Clubhsell, unknown; (5) Alabama Pigtoe, unknown; (6) Alvord Cutthroat Trout, 1920s; (7) 'Āmaui, 1860s; (8) American Chestnut Moth, unknown; (9) Amistad Gambusia, 1987; (10) Angled Riffleshell, 1967; (11) Antioch Dunes Shieldback Katydid, unknown; (12) Arc-form Pearly Mussel, 1940; (13) Arcuate Pearly Mussel, unknown; (14) Ash Meadows Killifish, 1948; (15) Bachman's Warbler, 1988; (16) Bigmouth Rocksnail, unknown; (17) Bishop's 'Ō'ō, 1980s; (18) Black Mamo, 1907; (19) Blackfin Cisco, 1969; (20) Blue Walleye, 1983; (21) Boulder Snail, unknown; (22) Brown Pigtoe, unknown; (23) Cahaba Pebblesnail, 1965; (24) California Golden Bear, 1922; (25) Carolina Elktoe, unknown; (26) Carolina Parakeet, 1918; (27) Catahoula Salamander, 1964; (28) Cascade Mountain Wolf, 1940; (29) Central Valley Grasshopper, unknown; (30) Channeled Pebblesnail, unknown; (31) Chestnut Casebearer Moth, 1900; (32) Chestnut Ermine Moth, unknown; (33) Clear Lake Splittail, 1970s; (34) Closed Elimia, 1967; (35) Cobble Elimia, unknown; (36) Constricted Elimia, unknown; (37) Coosa Elktoe, unknown; (38) Coosa Pigtoe, unknown; (39) Coosa Rocksnail, unknown; (40) Corded Purg, unknown; (41) Deepwater Cisco, 1952; (42) Dusky Seaside Sparrow, 1987; (43) Eastern Cougar, 2011; (44) Eastern Elk, 1887; (45) Eskimo Curlew, 1981; (46) Eelgrass Limpet, 1920s; (47) *Elimia gibbera*, unknown; (48) *Elimia lachrymal*, unknown; (49) *Elimia macglameriana*, unknown; (50) Excised Slitshell, unknown; (51) Fine-Rayed Pearly Mussel, unknown; (52) Fish Lake Physa, unknown; (53) Franklin Tree, unknown; (54) Fusiform Elimia, unknown; (55) Grass Valley Speckled Dace, 1938; (56) Greater 'Akialoa, 1969; (57) Greater 'Amakihi, 1904; (58) Great Auk, 1852; (59) Greater Koa Finch, 1896; (60) Goff's Pocket Gopher, 1955; (61) Gull Island Vole, 1897; (62) Hairlip Sucker, 1893; (63) Hawai'i 'Akialoa, 1940; (64) Hawai'i Mamo, 1898; (65) Hawai'i 'Ō'ō, 1930s; (66) Hawaiian Rail, 1890; (67) Hearty Elimia, unknown; (68) Heath Hen, 1932; (69) *Hemigrapsus estellinensis*, 1963; (70) High-Spired Elimia, unknown; (71) Independence Valley Tui Chub, 1970s; (72) Ivory-Billed Woodpecker, 1987; (73) Kakawahie, 1963; (74) Kāma'o, 1990s; (75) Kioea, 1860s; (76) Kona Grosbeak, 1894; (77) Labrador Duck, 1880; (78) Lāna'i Hookbill, 1918; (79) Las Vegas Dace, 1986; (80) Laysan 'Apapane, 1923; (81) Laysan Rail, 1944; (82) Lewis Pearly Mussel, unknown; (83) Lesser Koa Finch, 1891; (84) Lined Pocketbook, unknown; (85) Longjaw Cisco, 1975; (86) Maryland Darter, 1988; (87) Merriam's Elk, 1906; (88) *Moho braccatus*, 1987; (89) Navassa Curly-Tailed Lizard, 1970; (90) Navassa Island Dwarf Boa, late 1800s; (91) Navassa Island Iguana, late 1800s; (92) Nearby Pearly Mussel, 1901; (93) New Mexico Sharp-Tailed Grouse, 1952; (94) Nukupu'u, 2000; (95) O'ahu 'Alauahio, 1990s; (96) O'ahu 'Ō'ō, 1860s; (97) Ochlockonee Arcmussel, unknown; (98) Oloma'o, 1980s; (99) Pagoda Slitshell, unknown; (100) Pahranagat Spinedace, unknown;(101) Pallid Beach Mouse, 1959; (102) Pasadena Freshwater Shrimp, 1933; (103) Passenger Pigeon, 1914; (104) Pecatonica River Mayfly, unknown; (105) Phantom Shiner, 1975; (106) Po'o-uli, 2004; (107) Pupa Elimia, unknown; (108) Pygmy Elimia, unknown; (109) Pyramid Slitshell, unknown; (110) Raycraft Ranch Killifish, unknown; (111) Ribbed Elimia, unknown; (112) Ribbed Slitshell, unknown;

(113) Robert's Stonefly, unknown; (114) Rocky Mountain Locust, 1902; (115) Rubious Cave Amphipod, unknown; (116) Rough-Lined Elimia, unknown; (117) Round Slitshell, unknown; (118) Sampson's Pearly Mussel, unknown; (119) Sandhills Crayfish, unknown; (120) Sea Mink, 1860; (121) San Marcos Gambusia, 1983; (122) Shortnose Cisco, 1985; (123) Shoal Sprite, unknown; (124) Short-Spired Elimia, unknown; (125) Silvernose Trout, 1930; (126) Sloane's Urania Butterfly, 1894; (127) Smith Island Cottontail, 1987; (128) Snake River Sucker, unknown; (129) Sooty Crayfish, late 1800s; (130) Southern Rocky Mountain Wolf, 1935; (131) Steward's Pearly Mussel, unknown; (132) Striate Slitshell, unknown; (133) Tacoma Pocket Gopher, 1970; (134) Tecopa Pupfish, 1979; (135) Thicktail Chub, 1950s; (136) *Thismia americana*, 1916; (137) Turgid-Blossom Pearly Mussel, unknown; (138) 'Ula-'ai-hawane, 1937; (139) Umbilicate Pebblesnail, unknown; (140) Utah Lake Sculpin, 1928; (141) Yellowfin Cutthroat Trout, 1903; (142) Xerces Blue Butterfly, 1943.

6. Unfortunately, this does not mean that all species can live on these lands. This is important because even if conservation policies are effective, they may have little or no impact on endangered species because the habitats on those lands may be irrelevant to the survival of a species. We thank Ragnhild Sollund for bringing this issue to our attention.

References

Aatola, E. (2012) 'Differing philosophies: Criminalization and the stop the Huntingdon animal cruelty debate'. In Ellefsen, R., Sollund, R. and Larsen, G. (eds) *Eco-global Crimes: Contemporary Problems and Future Challenges*. Oxon, UK: Ashgate, pp. 157–180.

Ando, A., Camm, J., Polasky, S. and Solow, S. (1998) 'Species distributions, land values, and efficient conservation', *Science*, 279(5359): 2126–2128.

Barnosky, A.D., Matzke, N., Tomiya, S., Wogan, G.O.U., Swartz, B., Quental, T.B., Marshall, C., McGuire, J.L., Lindsey, E.L., Maguire, K.C., Mersey, B. and Ferrer, E.A. (2011) 'Has the Earth's sixth mass extinction already arrived?' *Nature*, 471(7336): 51–57.

Beirne, P. (2009) *Confronting Animal Abuse: Law, Criminology, and Human-animal Relationships*. Lanham, MD: Rowman and Littlefield Publishers.

Beirne, P. (1999) 'For a nonspeciesist criminology: Animal abuse as an object of study', *Criminology*, 37(1): 117–148.

Bjørkdahl, K. (2012) 'The rhetorical meaning of a crime called speciesism: The reception of animal liberation'. In Ellefsen, R., Sollund, R. and Larsen, G. (eds) *Eco-global Crimes: Contemporary Problems and Future Challenges*. Oxon, UK: Ashgate, pp. 71–90.

Boekhout van Solinge, T. (2008) 'Eco-crime: The tropical timber trade'. In Siegel, D. and Nelen, H. (eds) *Organized Crime: Culture, Markets and Policies*. New York: Springer, pp. 97–111.

Brewer, T.D., Cinner, J.E., Fisher, R., Green, A. and Wilson, S.K. (2012) 'Market access, population density, and socioeconomic development explain diversity and functional group biomass of coral reef fish assemblages', *Global Environmental Change*, 22(2): 399–406.

Brooks, T.M., Pimm, S.L. and Oyugi, J.O. (1999) 'Time lag between deforestation and bird extinction in tropical forest fragments', *Conservation Biology*, 3(5): 1140–1150.

Brown, R.M., and Laband, D.N. (2006) 'Species imperilment and spatial patterns of development in the United States', *Conservation Biology*, 20(1): 239–244.

Burkett, P. and Foster, J.B. (2006) 'Metabolism, energy, and entropy in Marx's critique of political economy: Beyond the Podolinsky myth', *Theory and Society*, 35(1): 109–156.

Burns, Thomas J., Kick, Edward L., Murray, David A., and Murray, Dixie A. (1994). 'Demography, development and deforestation in a world-system perspective', *International Journal of Comparative Sociology*, 35(3–4): 221–239.

Carson, R. (2002) *Silent Spring*. Boston: Houghton Mifflin Harcourt.

Chundawat, R.S., Habib, B., Karanth, U., Kawanishi, K., Ahmad Khan, J., Lynam, T., Miquelle, D., Nyhus, P., Sunarto, S., Tilson, R. and Wang, S. (2013) *Panthera Tigris*, The IUCN Red List of Threatened Species. Version 2014.3. Accessed on 9 March 2015 at http://www.iucnredlist.org/details/15955/0.

Clifford, M. (ed.) (1998) *Environmental Crime: Enforcement, Policy, and Social Responsibility*. Burlington, MA: Jones and Bartlett Learning.

Clark, B. and York, R. (2005) 'Carbon metabolism: Global capitalism, climate change, and the biospheric rift', *Theory and Society*, 34(4): 391–428.

Clarke, R.V. and Rolf, A. (2013) 'Poaching, habitat loss and the decline of neotropical parrots: A comparative spatial analysis', *Journal of Experimental Criminology*, 9: 333–353.

Clausen, R. and Clark, B. (2005) 'The metabolic rift and marine ecology: An analysis of the oceanic crisis within capitalist production', *Organization and Environment*, 18(4): 422–444.

Clausen, R. and York, R. (2008) 'Global biodiversity decline of marine and freshwater fish: A cross-national analysis of economic, demographic, and ecological influences', *Social Science Research*, 37(4): 1310–1320.

Colborn, T., Dumanoski, D. and Myers, J.P. (1996) *Our Stolen Future: Are we Threatening our Fertility, Intelligence, and Survival?* London: Little, Brown.

Czech, B. (2000) 'Economic growth as the limiting factor for wildlife conservation', *Wildlife Society Bulletin*, 28: 4–14.

Czech, B. and Krausman, P.R. (1997) 'Distribution and causation of species endangerment in the United States', *Science*, 277(5329): 1116–1117.

Czech, B., Krausman, P.R. and Devers, P.K. (2000) 'Economic associations among causes of species endangerment in the United States', *BioScience*, 50(7): 593–601.

Dinerstein, E., Loucks, C., Wikramanayake, E., Ginsberg, J., Sanderson, E., Seidensticker, J., Forrest, J., Bryja, G., Heydlauff, A., Klenzendork, S., Leimbruger, P., Mills, J., O'Brien, T., Shrestha, M., Simon, R. and Songer, M. (2007) 'The fate of wild tigers', *Bioscience*, 57(6): 508–514.

Dobson, A.P., Rodriguez, J.P., Roberts, W.M. and Wilcove, D.S. (1997) 'Geographic distribution of endangered species in the United States', *Science*, 275(5299): 550–553.

Eliason, S. (2012) 'From the king's deer to a capitalist commodity: A social historical analysis of the poaching law', *International Journal of Comparative and Applied Criminal Justice*, 36(2): 133–148.

Ehrlich, P.R. and Ehrlich, A.H. (1996) *Betrayal of Science and Reason: How Anti-Environmental Rhetoric Threatens Our Future.* Washington, D.C.: Island Press.

Feuerer, T. and Hawksworth, D.L. (2007) 'Biodiversity of lichens, including a world-wide analysis of checklist data based on Takhtajan's floristic regions', *Biodiversity and Conservation*, 16(1): 85–98.

Foster, J.B. (1992) 'The absolute general law of environmental degradation under capitalism', *Capitalism, Nature, Socialism*, 3(3): 77–82.

Foster, J.B. (1994) *The Vulnerable Planet: A Short Economic History of the Environment.* New York: Monthly Review Press.

Foster, J.B. (1997) 'The crisis of the Earth: Marx's theory of ecological sustainability as a nature-imposed necessity for human production', *Organization & Environment*, 10(3): 278–295.

Foster, J.B. (1999) 'Marx's theory of metabolic rift: Classical foundations for environmental sociology', *American Journal of Sociology*, 105(2): 366–405.

Foster, J.B. (2000) *Marx's Ecology: Materialism and Nature.* New York: Monthly Review Press.

Foster, J.B. and Clark, B. (2004) 'Ecological imperialism: The curse of capitalism', *Socialist Register*, 40: 186–201.

Foster, J.B., Clark, B. and York, R. (2010) *The Ecological Rift: Capitalism's War on the Earth.* New York: Monthly Review Press.

Franz, A. (2011) 'Crimes against water: Non-enforcement of state water pollution laws', *Crime, Law and Social Change*, 56(1): 27–51.

Gibbon, J.W., Scott, D.E., Ryan, T.J., Buhlmann, K.A., Tuberville, T.D., Metts, B.S., Greene, J.L., Mills, T., Leiden, Y., Poppy, S. and Winnie, C.T. (2000) 'The global decline of reptiles, déjà vu amphibians', *BioScience*, 50(8): 653–666.

Gould, K.A., Pellow, D.N. and Schnaiberg, A. (2008) *The Treadmill of Production: Injustice and Unsustainability in the Global Economy.* Boulder: Paradigm.

Greife, M.B. and Stretesky, P.B. (2013) 'Treadmill of production and state variations in civil and criminal liability for oil discharges in navigable waters.' In South, N. and Brisman, A. (eds). *Routledge International Handbook of Green Criminology.* London: Routledge, pp. 150–166.

Hagstedt, J. and Korsell, L. (2012) 'Unlawful hunting of large carnivores in Sweden'. In Ellefsen, R., Sollund, R. and Larsen, G. (eds) *Eco-global Crimes: Contemporary Problems and Future Challenges.* Oxon, UK: Ashgate, pp. 209–232.

Hawksworth, D.L. (2001) 'The magnitude of fungal diversity: The 1.5 million species estimate revisited', *Mycological Research*, 105(12): 1422–1432.

Hornborg, A. (1998) 'Towards an ecological theory of unequal exchange: Articulating world system theory and ecological economics', *Ecological Economics*, 25(1): 127–136.

Hoffmann, J.P. (2004) 'Social and environmental influences on endangered species: A cross-national study', *Sociological Perspectives*, 47(1): 79–107.

IUCN (International Union of Conservation of Nature) 2013, *IUCN Red* List. Accessed on 1 December 2013 at http://www.iucnredlist.org/documents/2001CatsCrit_Summary_EN.pdf.

International Union for the Conservation of Nature (2013) *Table 1: Number of Threatened Species by Major Groups of Organisms (1996–2013).* Accessed on 1 December 2013 at http://cmsdocs.s3.amazonaws.com/summarystats/2013_2_RL_Stats_Table1.pdf.

Jorgenson, A.K. (2003) 'Consumption and environmental degradation: A cross-national analysis of the ecological footprint', *Social Problems*, 50(3): 374–394.

Jorgenson, A.K. (2004) 'Uneven processes and environmental degradation in the world-economy', *Human Ecology Review*, 11(2): 103–117.

Jorgenson, A.K. (2006) 'Unequal ecological exchange and environmental degradation: A theoretical proposition and cross-national study of deforestation, 1990–2000', *Rural Sociology*, 71(4): 685–712.

Jorgenson, A.K. and Clark, B. (2011) 'Societies consuming nature: A panel study of the ecological footprints of nations, 1960–2003', *Social Science Research*, 40(1): 226–244.

Karanath, K.U. and Stith, B.M. (1999) 'Prey depletion as a critical determinant of tiger population viability'. In Seidensticker, J. Christie, S. and Jackson, P. (eds) *Riding the Tiger: Tiger Conservation in Human-Dominated Landscapes*. Cambridge: Cambridge University Press, pp. 100–113.

Kenney, J.S., Smith, J.L.D., Starfield, A.M. and McDougal, C.W. (1995) 'The long-term effects of tiger poaching on population viability', *Conservation Biology*, 9(5): 1127–1133.

Kirkland Jr. G.L. and Ostfeld, R.S. (1999) 'Factors influencing variation among states in the number of federally listed mammals in the United States', *Journal of Mammalogy*, 80(3): 711–719.

Lawson, K. and Vines, A. (2014) *Global Impacts of the Illegal Wildlife Trade: The Cost of Crime, Insecurity and Institutional Erosion*. London: Chatham House/The Royal Institute of International Affairs.

Linkie, M., Chapron, G., Martyr, D.J., Holden, J. and Leader-Williams, N. (2006) 'Assessing the viability of tiger subpopulations in a fragmented landscape', *Journal of Applied Ecology*, 43(3): 576–586.

Lomolino, M.V., Channell, R., Perault, D.R. and Smith, G.A. (2001) 'Downsizing nature: Anthropogenic dwarfing of species and ecosystems'. In Lockwood, J. and McKinney, M. (eds) *Biotic Homogenization*. New York: Springer, pp. 223–243.

Long, M.A., Stretesky, P.B. and Lynch, M.J. (2014) 'The treadmill of production, planetary boundaries, and green criminology'. In Spapens, T., Kluin, M. and White, R. (eds) *Environmental Crime and Its Victims: Perspectives Within Green Criminology*. Farnham, UK: Ashgate, forthcoming.

Luck, G.W. (2007) 'A review of the relationships between human population density and biodiversity', *Biological Reviews*, 82(4): 607–645.

Lynch, M.J., Long, M.A., Barrett, K.L. and Stretesky, P.B. (2013) 'Is it a crime to produce ecological disorganization? Why green criminology and political economy matter in the analysis of global ecological harms', *British Journal of Criminology*, 55(3): 997–1016.

McKinney, L.A., Kick, E.L. and Fulkerson, G.M. (2010), 'World system, anthropogenic, and ecological threats to bird and mammal species: A structural equation analysis of biodiversity loss', *Organization and Environment*, 23(1): 3–31.

Naidoo, R. and Adamowicz, W.L. (2001) 'Effects of economic prosperity on numbers of threatened species', *Conservation Biology*, 15(4): 1021–1029.

Ngoc, A.C. and Wyatt, T. (2013) 'A green criminological exploration of illegal wildlife trade in Vietnam', *Asian Journal of Criminology*, 8(2): 129–142.

Nurse, A. (2013) *Animal Harm: Perspectives on Why People Harm and Kill Animals.* Oxon, UK: Ashgate..

Orme, C.D.L., Davies, R.G., Burgess, M., Eigenbrod, F., Pickup, N., Olson, V.A., Webster, A.J., Ding, T.-S., Rasmussen, P.C., Ridgely, R.S., Stattersfield, A.J., Bennett, P.M., Blackburn, T.M., Gaston, K.J. and Owens, I.P.F. (2005) 'Global hotspots of species richness are not congruent with endemism or threat', *Nature*, 436(7053): 1016–1019.

Parmesan, C. and Yohe, G. (2003) 'A globally coherent fingerprint of climate change impacts across natural systems', *Nature*, 421(6918): 37–42.

Pellow, D. (2004) 'The politics of illegal dumping: An environmental justice framework', *Qualitative Sociology*, 27(4): 511–525.

Pires, S.F. and Clarke, R.V. (2012) 'Are parrots CRAVED? An analysis of parrot poaching in Mexico', *Journal of Research in Crime and Delinquency*, 490(1): 122–146.

Pimm, S.L., and Raven, P. (2000) 'Biodiversity: Extinction by numbers', *Nature*, 403(6772): 843–845.

Pires, S.F. and Moreto, W.V. (2011) 'Preventing wildlife crimes: Solutions that can overcome the tragedy of the Commons', *European Journal on Criminal Policy and Research*, 17(2): 101–123.

Rockstron, J., Steffen, W., Noone, K., Perrson, A., Chapin, F.S., Lambin, E., Lenton, T.M., Scheffer, M., Folke, C., Schellnhuber, H.J., Nykvst, B., de Wit, C.A., Hughes, T., van der Leeuw, S., Rodhe, H., Sorlin, S., Snyder, P.K., Costanza, R. Svedin, U., Falekmark, M., Karlberg, L., Corell, R.W., Fabry, V.J., Hanson, J., Walker, B., Liverman, D., Richardson, K., Crutzen, P. and Foley, J. (2009) 'Planetary boundaries: Exploring the safe operating space for humanity', *Ecology and Society*, 14(2): 32–47.

Ruggiero, V. (2013) *The Crimes of the Economy: A Criminological Analysis of Economic Thought.* London: Routledge.

Ruggiero, V. and South, N. (2010) 'Critical criminology and crimes against the environment', *Critical Criminology*, 18(4): 245–250.

Santos-Paulino, A.U. (2012) 'Trade, income distribution and poverty in developing countries: A Survey'. Paper # 207, United Nations Conference on Trade and Development. Accessed on 1 August 2014 at http://ideas.repec.org/p/unc/dispap/207.html.

Schnaiberg, A. (1980) *The Environment: From Surplus to Scarcity.* New York: Oxford University Press.

Schnaiberg A and Gould, K.A. (1994) *Environment and society: The enduring conflict.* New York: St Martin.

Shandra, J.M., Leckband, C., McKinney, L.A. and London, B. (2008) 'Ecologically unequal exchange, world polity, and biodiversity loss: A cross-national analysis of threatened mammals', *International Journal of Comparative Sociology*, 50(3–4): 285–310.

Sekercioglu, C.H., Schneider, S.H., Fay, J.P. and Loarie, S.R. (2008) 'Climate change, elevational range shifts, and bird extinctions', *Conservation Biology*, 22(1): 140–150.

Smith, J.L., David, Ahern, S.C. and McDougal, C. (1998) 'Landscape analysis of tiger distribution and habitat in Nepal', *Conservation Biology*, 12(6): 1338–1346.

Sollund, R. (2013) 'Animal trafficking and trade: Abuse and species injustice,' In Westerhuis, D., Walters, R. and Wyatt, T. (eds) *Emerging Issues in Green Criminology: Exploring Power, Justice and Harm.* Palgrave, pp. 72–90.

Sollund, R. (2008) 'Causes for speciesism: Difference, distance and denial'. In Sollund, R. (ed.) *Global Harms. Speciesism and Ecological Crime.* New York: Nova, pp. 109–130.

Sollund, R. (2011) 'Expressions of speciesism: The effects of keeping companion animals on animal abuse, animal trafficking and species decline,' *Crime, Law and Social Change,* 55(5): 437–451.

Sollund, R. (2012a) 'Speciesism as doxic practice versus valuing different practices'. In Ellefsen, R., Sollund, R. and Larsen, G. (eds) *Eco-global Crimes: Contemporary Problems and Future Challenges.* Oxon, UK: Ashgate, pp. 91–114.

Sollund, R. (2012b) 'Oil production, climate change and species decline: The case of Norway'. In White, R. (ed.) *Climate Change from a Criminological Perspective.* New York: Springer, pp. 135–147.

Sollund, R. (2013) 'Animal trafficking and trade: Abuse and species injustice', In Westerhuis, D., Walters, R. and Wyatt, T. (eds) *Emerging Issues in Green Criminology: Exploring Power, Justice and Harm.* Palgrave, pp. 72–90.

Steffen, W., Crutzen, P.J. and McNeill, J.R. (2007) 'The anthropocene: Are humans now overwhelming the great forces of nature', *Ambio: A Journal of the Human Environment,* 36 (8): 614–621.

Steffen, W., Grinevald, J., Crutzen, P. and McNeill, J. (2011) 'The Anthropocene: Conceptual and historical perspectives', *Philosophical Transactions of the Royal Society A: Mathematical, Physical and Engineering Sciences,* 369: 842–867.

Steingraber, S. (1997) *Living Downstream: A Scientist's Personal Investigation of Cancer and the Environment.* New York: Vintage Books.

Stork, N.E. (2010) 'Re-assessing current extinction rates', *Biodiversity and Conservation,* 19(2): 357–371.

Stretesky, P.B., Long, M.A. and Lynch, M.J. (2013) *The Treadmill of Crime: Political Economy and Green Criminology.* London, Routledge.

Stuart, S.N., Chanson, J.S., Cox, N.A., Young, B.E., Rodrigues, A.S.L., Fischman, D.L. and Waller, R.W. (2004) 'Status and trends of amphibian declines and extinctions worldwide', *Science,* 306(5702): 1783–1786.

Svard, P.-A. (2012) 'The ideological fantasy of animal welfare: A Lacanian perspective on the reproduction of speciesism', In Ellefsen, R., Sollund, R. and Larsen, G. (eds) *Eco-global Crimes: Contemporary Problems and Future Challenges.* UK: Ashgate, Oxon, pp. 115–132.

Tabb, W. and Sawers, L. (1984) *Marxism and the Metropolis: New Perspectives in Urban Political Economy.* New York: Oxford University Press.

Thomas, C.D., Cameron, A., Green, R.E., Bakkenes, M., Beaumont, L.J., Collingham, Y.C., Erasmus, B.F.N., de Siqueira, M.F., Grainger, A., Hannah, L., Hughes, L., Huntley, B., van Jaarsveld, A.S., Midgley, G.F., Miles, L., Ortega-Huerta, M.A., Peterson, A.T., Phillips, O.L. and Williams, S.E. (2003) 'Extinction risk from climate change', *Nature,* 427(6970): 145–148.

Thuiller, W., Lavorel, S., Araújo, M.B., Sykes, M.T. and Prentice, I.C. (2005) 'Climate change threats to plant diversity in Europe', *Proceedings of the National Academy of Sciences of the United States of America,* 102(23): 8245–8250.

Tillman, D.A. (2012) *Wood as an Energy Resource.* Amsterdam: Elsevier.

USFWS (United States Fish and Wildlife Service) 2013, *Endangered List.* Accessed on 30 November 2013 at http://www.fws.gov/endangered/map/index.html.

Vačkář, D. (2012) 'Ecological footprint, environmental performance and biodiversity: A cross-national comparison', *Ecological Indicators,* 16: 40–46.

Vail, B. (2009) 'Municipal waste management policy in Europe: How the treadmill of production undermines sustainability goals', *International Journal of Sustainable Society*, 1(3): 224–239.

van Solinge, T.B. (2010) 'Equatorial deforestation as a harmful practice and a criminological issue.' In White, R. (ed.) *Global Environmental Harm: Criminological Perspectives*. Devon: Willan, pp. 20–36.

Wackernagel, M. and William, R. (2012) *Our Ecological Footprint: Reducing Human Impact on the Rarth*. Canada: Gabriola Island.

Walters, R. (2010) *Eco Crime and Genetically Modified Food*. New York: Routledge.

Wargo, J. (1998) *Our Children's Toxic Legacy: How Science and Law Fail to Protect Us from Pesticides*. Newhaven, Conn: Yale University Press.

Ween, G.B. (2012) 'Enacting human and non-human indigenous: Salmon, Sami and Norwegian natural resource management'. In Ellefsen, R., Sollund, R. and Larsen, G. (eds) *Eco-global Crimes: Contemporary Problems and Future Challenges*. Oxon, UK: Ashgate, pp. 295–312.

Wilcove, D.S. and Master, L.L. (2005) 'How many endangered species are there in the United States?' *Frontiers in Ecology and the Environment*, 3(8): 414–420.

Wilcove, D.S., Rothstein, D., Dubow, J., Phillips, A. and Losos, E. (1998) 'Quantifying threats to imperiled species in the United States', *BioScience*, 48(8): 607–615.

Woodroffe, R. (2000) 'Predators and people: Using human densities to interpret declines of large carnivores,' *Animal Conservation*, 3(2): 165–173.

Wyatt, T. (2009) 'Exploring the organization of Russia far east's illegal wildlife trade: Two case studies of the illegal fur and illegal falcon trades', *Global Crime*, 10(1–2): 144–154.

Wyatt, T. (2011) 'The illegal trade of raptors in the Russian Federation', *Contemporary Justice Review*, 14(2): 103–123.

Wyatt, T. (2013) *Wildlife Trafficking: A Deconstruction of the Crime, the Victims, and the Offenders*. Basingstoke: Palgrave Macmillan.

Zalasiewicz, J., Williams, M., Steffen, W. and Crutzen, P. (2010) 'The new world of the Anthropocene', *Environmental Science & Technology*, 44(7): 2228–2231.

8

The Illegal Wildlife Trade from a Norwegian Outlook: Tendencies in Practices and Law Enforcement

Ragnhild Aslaug Sollund

The illegal trade in free-born[1] animals and products made from their bodies is one of the fastest-growing illegal trades worldwide (e.g. European Commission, 2014). Usually the term 'wildlife'[2] is intended to include plants, but in this chapter my focus is on animals. The so-called 'wildlife trade' is repeatedly positioned among the drugs trade, illegal arms trade and human trafficking (e.g. Warchol, 2007; Zimmerman, 2003; South and Wyatt, 2011). The trade in endangered species is regulated under the CITES convention, which Norway signed in 1976. Some 5,000 animal species are now listed as threatened in the CITES appendices (I, II and III), many because of trade, or the combination of loss of habitat and trade (WWF, 2014a; Reid, 1992). The convention provides a framework that is to be respected by each party, which have to adopt their own domestic legislation to ensure that CITES is implemented at national level.[3] There are now 180 member parties. It is important to note that trade in free-born animals (or CITES-listed animals bred in captivity) is not criminalized *per se*; rather, the regulation through CITES implies that individuals of most species may legally be subject to abduction, trafficking and/or killing, while individuals of other species are protected depending on their degree of endangerment (Sollund, 2011).

My intention with this chapter is to see what practices with regard to the free-born animal trade are present in Norway through the examination of data collected for an ongoing research project.[4] Further, I will discuss whether law enforcement of these crimes are adequate and efficient. Towards the end of the chapter I will discuss my findings from the perspective of green criminology – more specifically in relation to the field of green criminology that Rob White has referred to as speciesist criminology (2013: 25). A better name for the field would be anti-specieisist criminology, due to its non-anthropocentric inclusion of non-human animal harm and the focus on species justice. Given the topic of the research – the international and transnational trade in free-born animals – this field is necessarily also eco-global and environmental (White, 2011, 2013).

Data sources

My project is based on qualitative methods. I have carried out 17 interviews[5] in Norway. I have 46 confiscation reports from customs of animals and animal parts which have been detected either through mail delivery or at borders, in addition to confiscation reports which appear in police files. The attorney general in Norway has further permitted me an insight into a large number of criminal cases related to the smuggling of animals and animal products to Norway, and cases of plundering nests and killing protected predator birds in Norway, altogether 700–800 cases. Offenders who feature in these files are usually convicted of other crimes, such as violence or drug-related offences, and are also found guilty of, say, keeping a reptile which may previously have been smuggled into the country (but often this is not revealed in the files). In these cases typically the Animal Welfare Act and the associated regulation against the importation and keeping of exotic species is applied. It is fairly rare that reference is made to the CITES regulation in my material from the police and in the court verdicts. Usually other regulations are applied without any mention of CITES. In many cases police and customs do not attempt to establish whether the animal is CITES listed or not. Reptiles are, for example, often referred to simply as snakes or lizards in the penal cases in which the breach of the Animal Welfare Act is subsumed under crimes that breach other laws with possible stricter punishment, most often connected to violence or drugs. In addition to this Norwegian material I have interview data[6] from Colombia and Brazil which includes 14 interviews with wildlife experts and veterinarians in Colombia, and 4 interviews from Brazil, including with the São Paolo Environmental Military Police.

Rather than providing a picture of total trafficking and smuggling attempts, the confiscation reports from customs may give an indication of its priorities regarding CITES. Many seizures appear to be random rather than the outcome of deliberate attempts to uncover

CITES-listed species, according to my interviews with customs. It has targets for drug confiscations but not for CITES, which make it prioritize looking for drugs rather than CITES-listed species (see also Runhovde, forthcoming).[7]

On this occasion I will discuss the punishment that is imposed on offenders in terms of its individual and general preventative effect, largely based on the Norwegian data. I will concentrate on the files and reports that I have accessed from the police and customs in addition to concluded verdicts from Lovdata,[8] the Norwegian verdict database where a large number of verdicts are available.

Norwegian legislation contains several regulations which prohibit the importation of live animals to Norway from abroad, among them Viltloven (Wildlife Law §47),[9] The Law of Biodiversity[10] and the Animal Welfare Act,[11] with accompanying regulations against the importation, trade and keeping of 'exotic species', with a maximum sentence of six months imprisonment.[12] Another law which is applied in CITES cases is the Law on the Regulation of Importation and Exportation from 6 June 1997 [Lov om innførsels og utførselsregulering][13] (maximum sentence six months imprisonment). This may have been used for both flora and fauna that are CITES listed.

CITES cases are not consistently coded, which makes it impossible to track them and get an overview even of the registered cases. They are registered as misdemeanours rather than crimes (Svae-Grotli, 2014). The police also variously use different laws and regulations in their reports and statistics.

The Norwegian CITES regulation of November 2002 nr. 1976[14] is, when regarding animals, based on Wildlife Law §56, with a maximum prison sentence of two years. The regulation was revised in 2013 and the new regulation is also based in the Law of Biodiversity §26. The maximum sentence was thus increased to up to three years. With the new regulation, breaches of CITES may be registered as crimes rather than misdemeanours (Svae-Grotlie, 2014).

Taking CITIES listed animals in and out of Norway is illegal unless the necessary permissions (CITES export and import certificates) are obtained from the Norwegian Environment Agency in advance. Whether permission is granted may be of less importance to the animal's welfare – being protected by legislation may, as we shall see, have an adverse effect. This is one reason why I adopt a more holistic, green criminology perspective in which I acknowledge that being trafficked whether in accordance with or against a law or regulation is harmful and unjust and breaches both of individual animal rights, species – and ecological justice (Sollund, 2013; White, 2013).

Before I turn to the picture which appears regarding what animals and animal products are trafficked to Norway and how enfringements of CITES are reacted to by the police and the judicial system, I will give a brief overview of selected and relevant typologies of wildlife trafficking, in terms of species and motivations, as an introduction to the findings from the case study which shows that Norway, although geographically on the outskirts of Europe, is indeed part of a globalized world.

The international picture of illegal wildlife trade

Globalization in terms of the movement of people, of the shared Internet and thus also of information about the supposed value of animal products and where to get them, whether medicinal items or collectors' chosen products such as ivory, accounts for much of the increase in the legal and illegal wildlife trade (IFAW, 2008). Migrants do not leave their cultural (superstitious) practices in their country or place of origin behind but bring them to their new place of residence. In Colombia and Brazil this phenomenon is also evident in the practices of pet-keeping. There are long traditions of keeping parrots, monkeys and other free-born animals in Latin America (e.g. Guzman et al., 2007; Drews, 2002; Zimmerman, 2003; Weston and Menon, 2009). These are particularly widespread practices among the lower strata of the population, who during recent decades have moved into the cities (Traffic, interview). Through this movement, which is also taking place in the rest of Latin America, they bring these practices with them and also provide a market for domestic trafficking.

The parrot trade

Parrots are trafficked to internal and international markets (Weston and Menon, 2009; Gonzales, 2003; Herrera and Hennessey, 2007; Eniang et al., 2008; Metz, 2007; Guzman et al., 2007). The trade in wild parrots threatens a large number of species, although the greatest threat to many is habitat loss – for example, due to logging and mining, according to my data from Colombia with Traffic and Proaves (a bird conservation organization based in Colombia).[15] The problem for many parrot species is that they are under pressure from many different directions. According to du Plessis (2000: 21), of 90 species that are at risk of global extinction, 81% are threatened by habitat loss or degradation,

43% by trade in live birds, with at least 31% threatened by both. A report on the parrot trade in Mexico documents that many of the 22 parrot species there are under severe pressure because of the trade through which the majority are trafficked domestically (Guzman, 2007: 94). Gastañaga et al. (2011: 76–77) say (with reference to Birdlife and IUCN) that illegal trade is thought to contribute to the threat to 66 parrot species worldwide, including 27 in South America, where it is also believed to have caused the extinction of Spix's Macaw in the wild. The mortality rates are extremely high and it is estimated that 75% die before reaching a purchaser, including during capture (Guzman et al., 2007). This means that in Mexico alone, 50,000–60,000 parrots die each year as consequence of the trade. An average of 31% die during transport, so traffickers rely on large volumes to cater for the high mortality rates, e.g. when birds drown when they are soaked under water to prevent them from screaming when traffickers face police controls (Weston and Menon, 2009). However, an interview with Proaves suggests that the mortality rates are even higher, up to 90%, and Guzman et al. (2007) emphasize that the death of eggs and nestlings because their parents have been captured is also part of the picture of depletion. It is estimated that in Mexico, 65,000 to 75,000 parrots are captured each year. My interviews from Colombia and Brazil, together with research on the parrot trade in Latin America (Weston and Menon, 2009; Castañaga et al., 2011; Guzman, 2007; Gonzales, 2003; Drews, 2002), Africa (Eniang et al., 2008) and many other places (Wright et al., 2001), show that the abduction of parrots from their nests, usually when they are small, is a widespread practice with long cultural roots (parrots have been traded and kept as pets for thousands of years (Weston and Menon, 2009), and that it represents a huge threat to several species because up to 70% of nestlings of some species are abducted (Wright et al., 2001).

In more recent times, robbing nests and trade the abducted birds has been perceived as a way of getting extra income in Latin America (e.g. Zimmerman, 2003; Gonzales, 2003), and the need to make extra income has for many been increased as a result of pressures to integrate into a market economy (Junglevagt for Amazonas, 1995 in Gonzales. 2003: 439). In Latin America, most parrots who are abducted are sold in local markets (Weston and Menon 2009; Gonzales, 2003; Guzman et al., 2007), and this practice continues even though it is common knowledge that it is illegal. One example of how this takes place was documented in Bogotá in 2013.[16] Questioned whether he had parrots for sale, the owner of a pet store said that they could be purchased in department stores in a block adjacent to Plaza de Restrepo. He suggested that he could inquire discretely there because 'it is forbidden and the sale of parrots is therefore clandestine'. A quick Google search with the question 'Where can I buy a parrot in Bogotá' confirmed that this was well known, and it showed that taking and keeping parrots from the forests is widely accepted (Weston and Menon, 2009). The birds are given inadequate food, are kept in small cages and, when they die prematurely, they are replaced (interview data from Colombia, Weston and Menon, 2009).

Animals used as ingredients for medicinal (and superstitious) reasons

The species which currently receive most attention as victims of trade that is also motivated by the presumed medicinal effect of the animal's byproducts is the rhinoceros. This attention is likely due to the degree of endangerment, the ferocity of the killers (heavily armed subversive groups often engage in this to acquire money for arms and to finance guerrilla wars, as in Sudan (Wyatt, 2013a) and the symbolic value that the species holds as representative of the 'wild' and being one of the 'big five'. Rhino horn is assumed to cure a large number of diseases, from cancer to hangovers, despite its makeup being the same as fingernails. Rhino horn is used in drinks among the Vietnamese upper class and also simply to display wealth (Minnaar, 2014; Vecchiatto, 2014). The growth of the Chinese and Vietnamese middle classes has led to an increased demand for rhino horn. Likewise tigers are in demand because, according to cultural superstitions, they are regarded as holding special capacities as talismans and also are used in traditional medicine (Emslie in Warchol, 2007; Lee, 1996). There remain only 3,200 tigers in the wild (while there are 5,000 in captivity in the USA alone, according to the WWF (2014). Black bears' gall bladders are used for superstitious medicinal purposes (among others) in many Asian countries and their trafficking entailed estimated seizures of both 2,801 bears or bear products between 2000 and 2011 according to a report by Traffic (Burgess et al., 2014: 14).

The reptile trade

To be used as medicine is also the fate of many reptile species (e.g. Minnaar, 2013; Ellis, 2005; Zhang et al., 2008; Alves et al., 2008; Santana, 2008), which are used widely in Brazil (Alves et al., 2008) and China (Zhang et al., 2008).

In my Norwegian data the reptiles that figure as confiscated are trafficked to become 'pets'.[17] This increasing trade is threatening many reptile species, such as in South East Asia (Foley, 2013; Australia Herbig,

2010). Trafficking in reptiles for the pet trade also has a huge impact because it endangers many species and is going on in the USA and Europe on a large scale, not least with turtles and tortoises. In fact, it is only in Iceland and Norway within Europe that keeping reptiles as pets is criminalized (Sollund, 2013). Further, reptiles are trafficked as skins, a typology of the trade which is also represented in several of the Norwegian confiscation reports.

Lots of reptiles are also trafficked to be eaten. In Colombia, for example, the rare Icotea Turtles are abducted in great numbers and brutally killed because they are 'white meat' rather than 'red meat', which is not eaten during Easter for cultural/religious reasons.

Typologies of wildlife trafficking to Norway

Norway's position is very much that of being a receiving country for wildlife, with the exception of birds, which may be trafficked out of the country or locally, as eggs (Miljøkrim, 2013). There are also other breaches of the Wildlife Law in Norway which involve the illegal killing of protected species, such as wolf, wolverine, lynx and brown bear, and also eagles and hawks, which may be killed or have their nests destroyed. These cases are often the outcome of human–animal conflict and will not be covered her (Sollund, 2015; Hagstedt and Korsell, 2012).

Parrot profit cases

I will present various cases which have different kinds of motivation in order to discuss what measures would be effective in preventing such crimes. I start by presenting a case in which a man was convicted in December 2013 for parrot trafficking,[18] as well as alcohol and tobacco smuggling. This can be categorized as a profit case. The offender was smuggling parrots not to keep them as companion animals himself but as part of a business,[19] and this was not the first time. He was stopped by customs in 2010 with four African Grey Parrots, and a year later with eight birds of the same species. My CITES interviewee said that it was likely that these birds had been caught in the wild as there are large numbers of them coming into Europe,[20] and these are far cheaper than locally bred birds. The parrots had parasites which lends weight to this suggestion because hand-bred parrots are less likely to be infested.

The man was convicted for his crimes on 3 December 2013, more than three years after the first offence and two years after the last. The charge was made a year and a half after the last offence. Such a delay is typical of these cases. Since I received the confiscation reports I was in contact with three state prosecutors regarding this case, each said that they had recently got the case on their desk. When I called the last one who finally concluded the charges, he was unaware that the man had been caught before, and he was also unaware of previous related verdicts which could have been important in building the case. Interestingly, in this case I also found that for the first of these offences, in 2010, the offender was also fined NOK 25,000, but apparently he did not accept the fine since the case was brought to court (the police can impose a fine for a misdemeanour but if the fine is not accepted the case will be brought to court). The fine did not mention the birds, and in another document from customs in relation to the seizure, the birds were named as the wrong species, and their economic value was set at NOK 200 rather than NOK 5,000–7,000 each.

In court it was ruled that all of the parrots (plus hundreds of litres of alcohol and a kilo of tobacco) would be confiscated – or rather it was confirmed since the birds had already been killed. The argument for killing the four birds in 2010 was that they were 'only' on the CITES II list (less endangered than CITES I species). The offender was not expected to pay any administrative costs. He was given a 30-day prison sentence which was suspended for two years (so he was not imprisoned) and was ordered to pay a fine of NOK 5,000. In the verdict it stated that confiscation is necessary to ensure the effective enforcement of CITES. It was regarded as an aggravating circumstance by the judge that the man was convicted for breaches of the custom's law on several occasions and particularly so because trafficking was part of his business. However the judge who imposed the sentence was lenient owing to the long delay of the case in the judicial system.

This case shows clearly that the offender was wise when he (presumably) did not accept the first fine of NOK 25,000 NOK because it was subsequently reduced to only NOK 5,000 in Court. He also avoided going to prison as a result of the delay in the judicial system. In sum, one can say that this was mild punishment compared with the maximum sentence that the law allows of two years prison for the trafficking of the birds alone, disregarding the alcohol smuggling. It is interesting, though, that the argument for the confiscation of the parrots was given as their CITES II listing, upon which they were all subject to theriocide in 2010 (Beirne, 2014), and that this was done in order to ensure enforcement of CITES. However, the idea behind CITES is to protect endangered species, not to kill them, so it is hard to see that such enforcement corresponds with the intention behind the convention. In this case the birds were victims twice: first when abducted and trafficked (assuming they were indeed wild caught)[21] and then when they were killed in Norway.

To further illustrate how lenient the reactions are to these kinds of offences, another man who was stopped on the Svinesund border on 13 November 2011 with no less than 25 birds, many of which were CITES listed, received (and accepted) a fine of only NOK 10 000. It is reasonable to think that he didn't intend keeping all of the birds for himself and my data show that he too was a reoffender. In these parrot cases it seems that the prosecutors do not go to the trouble to look at previous CITES convictions. Such efforts would have allowed these cases to take precedence, such as the verdict from 1998 in which a man was convicted to six months prison for the trafficking of parrots and reptiles on many occasions (Sollund, 2013: 81). In that case the judge emphasized CITES and the need to respond strongly as a clear deterrent. The sentence and fine in 2014 are in comparison very lenient.

Souvenirs bought by tourists

A frequent category in my material is that of tourists who purchase products that are made from free-born animals. These usually qualify as what Wyatt (2013a) refers to as 'accidental offenders'. They may be unaware of the illegality of their purchase and the situation of the species involved. Norwegians travel a lot because most have the finances to do so. Thailand, for example, is a popular destination and is known to serve as an important source and transit country for wildlife trade.[22] Norwegian tourists typically smuggle items that are made of animal parts that are CITES listed (e.g. ivory, sea horses, jewellery made from sea turtles, black coral, feathers and shark teeth), either in their luggage or they have it sent by mail.

Common confiscations also include belts, purses, shoes, wallets, and liquor containing scorpions and snakes. For example, one case involved a family man smuggling two bottles containing two cobra snakes (*Naja atra* or *Naja kaouthia*; CITES II listed). He was fined NOK 12,000. Other offenders' confiscated animal products included pelts and heads of dead animals, one primate,[23] three Canadian Wolves, a Grizzly Bear, Black Bear pelts and a Sawfish snout (CITES I), for the last offender was fined NOK 3,000. Three other reports relate to stuffed predator birds, all of which were CITES listed. Other tourists who are aware of the rules may have items such as ivory figures or crocodile head trophies mailed to them in the regular post. A recent confiscation was of a stuffed Siberian Tiger (CITES I). Another way of purchasing the desired product is on the Internet, and one case that is currently under investigation involves the import of hundreds of ivory items as well as a large number of other species. This man involved said that he acted in a feverish way when participating in auctions on the Internet of ivory and other items, and that he lost track of the numbers of items that he had accrued in this way. The collecting of animal products (and dead animals) can become an obsession, as for collectors of other types of item (Nurse, 2013: 160).

The live reptile trade

There are several reports which concern tourists who bring live animals to Norway, typically a turtle or another reptile bought at a market, which is then trafficked in a pocket. Alternatively, reptiles are often trafficked to Norway in a more planned and deliberate way to circumvent the Norwegian ban on reptiles as pets (Sollund, 2013).

Among my data are a large number of penal cases – the 'combination cases' in which the reptile crimes are subsumed under other offences – which suggest that reptiles, especially snakes, are kept to attribute status to the owner in specific social circles where the use of drugs and other crime-related activities are part of the lifestyle. It appears from descriptions in police reports and verdicts about the conditions that the animals were kept in, and the circumstances in which they were bought, that, typically, little concern is given to the animals' welfare. A case awaiting trial in Norway at present shows that Norwegians are taking part in this international reptile trade, which not only abuses the individual reptiles and is a threat to their species, but is also terribly cruel and abusive to all of the rodents that are bred and are used as live food for the reptiles. In this case several people in Norway and Australia will, according to interviews, be charged with organized crime for breeding and trafficking a large number of reptiles for markets in Europe and Australia.

Animals as products for medicinal purposes

Three confiscation reports concern medicinal products that are made from near-extinct animal species: Leopard (CITES I), Tiger (CITES I) and Musk Deer (CITES II). All three offenders have a non-Norwegian background but live in Norway, and all were stopped at Oslo airport. The tiger bone case included ten boxes with ten pills in each, containing 5% tiger bone. The offender received a fine of NOK 4,500. In the case with the leopard bone, a species which according to Warchol et al. (2003) is taking over from the tiger as the source of medicinal products due to the scarcity of the tiger, the man received no punishment at all and the case was dismissed.

In the next section I will discuss whether the punishment which follows (or fails to follow) in these cases may be seen as efficient and adequate in order to deter individual offenders from reoffending, and to work as a general deterrent to prevent others from committing such crimes.

The consequences of the crimes for offenders

According to the eco-crime police overview of 87 cases that were categorized as the keeping and smuggling of exotic species, in the timespan of 2000–2010, 28 concluded with a fine and 31 were dismissed for different reasons, including because the cases had grown too old, which may be explained by their lack of priority. The police files show that a fine, irrespective of how the cases are coded, is the usual reaction to the trafficking of free-born animals to Norway, whether dead or alive. However, this is normally only NOK 5,000–10,000, so such a crime is thus punished as if it were a misdemeanour and equal to, or more lenient than, urinating in a public place.[24] To conclude the offences with a fine is the easiest way for a prosecutor to get a case off his desk, so it seems to be the preferred way in favour of initiating further investigation and bringing the case to court. Criticism has been made against the categorization of breaches of CITES as misdemeanours rather than crimes (Svae-Grotli, 2014: 33). Whether the police apply the Law of Importation or Exportation (Innførsels og utførselsloven §4) or the Wildlife Law (Viltloven §56), as mentioned, they dictate a maximum punishment of six months and two years, respectively. In none of the cases in my material was the maximum punishment applied. The categorization of breaches of CITES as misdemeanours probably contributes both to the lack of priority that these cases are given by the police and prosecutors, and also to the lack of esteem attached to the investigation of such crimes, which also partly explains their lack of priority (Svae-Grotli, 2014). As shown, this means that cases are shuffled from desk to desk, the prosecutors do not bother to go into previous cases to build up the current case, and usually the offender merely gets a fine. Further, many customs and police officers possess little knowledge about how to handle CITES cases, something which is clear from my material. For example, often the species to which a reptile belongs remains unidentified. A consequence of this level of punishment is that it has a minimal deterrent effect. For those who do this for profit, it is worth the risk – what they sell, say, African Grey Parrots for by far exceeds the fines imposed. The existence of threat of punishment for CITES crimes is only to a very small degree made known to the public (e.g. there are no advertisements about CITES at Norwegian airports, and the newspapers seldom write about it), so the possible punishment represents a poor general deterrent.

The consequences of the crimes for the direct victims

There is a general pattern in the cases where animals are found alive – either confiscated at the borders or when they are seized in private homes – that the animals involved are killed. In nearly all of the reptile cases that I have gone through, this is the outcome. In two of the 'combination cases' the police officers even state in the reports that they killed the animals themselves, but without giving details of how this was done. I strongly doubt that these killings have been in line with the Animal Welfare Act. This is also the outcome for most of the parrot-trafficking victims. In the parrot cases mentioned above, as shown, in the case of the four African Greys, it ended with tragedy. The other eight were sent to a zoo, Kristiansand dyrepark, as were most of the 25 birds that were trafficked by the other offender. In another case from 12 August 2014, six African Grey Parrots were killed, on 30 March 2014, eight parrots were killed because they exceeded the number that there were permits for.

One reason for this constant practice of theriocide of trafficking victims in Norway is that there are no facilities where they can be taken care of, unlike the situation in Colombia and Brazil, and at Heathrow Airport in the UK (Wyatt, 2013b). Some are sent to the zoo in Kristiansand and some are sent to the reptile park in Oslo, but these can take only a limited number and their interests in taking them in will depend on the species' rarity. Decisions about what to do with the animals are made after consulting with the Norwegian Environment Agency – the formal owner of all wildlife in Norway – where they will usually advise that the animals should be killed, especially if they are 'only' CITES II listed, as with the African Grey Parrot. This, in addition to the ways in which these cases generally suffer from a lack of priority in the judicial system, is another indication of the lack of political interest and priority that CITES has in Norway.

I find it extremely paradoxical that the legislation and regulations which are supposed to protect animals and animal species consistently seem to pay no attention to the animals themselves.[25] Animals are killed, usually by direct reference to the specific regulations, whether CITES, the Wildlife Law or the Animal Welfare Act. In the case of CITES this is not so surprising because the intention of the convention is to regulate trade in endangered species, so that the animals and plants that are listed continue to provide 'resources', for humans and continue to enrich our lives. As I have stated elsewhere (Sollund, 2011), CITES itself not only legitimates wildlife trade but encourages it. This is not least due to the normative signals that are conveyed by the convention that animals are merely part of nature, and that their value lies therein as resources. When the animals are usually killed when seized, it is clear that the regulation which should protect animals from trafficking and trade most often has the reverse effect for the victims involved.

It is also a huge problem that as long as trade in wildlife is legal, there are a large number of ways to circumvent the legislation through forgery of CITES permits, laundering and corruption (Warchol et al., 2003; Warchol, 2007; Lemieux and Clarke, 2009), in which illegal trade flourishes under the guise of being legal (Hutton and Webb, 2005; Sheperd et al., 2012; Rosen and Smith, 2010; Wyatt, 2013b). In one of the interviews carried out in Brazil, with the Chico Mendez Institute, it was also revealed that the foot rings that are used to identify bred parrots are used on wild parrots, often resulting in amputated legs. The São Paolo environmental police also gave examples of fraud, such as when a permit was given to take 100 fish but far more were taken. A recent confiscation report from Norway, also concerning parrots, showed the same way of circumventing the regulation – by taking in more parrots than there were CITES permits for. The 'surplus' birds were killed.

Even in the cases where the Animal Welfare Act is applied, it becomes evident that the focus of the prosecutors is not on the victims – the animals – but merely on the breach of law. In many ways this can be claimed to be the failure of the penal system, in which what counts is the law and breaches of law, rather than the victims of the offences. This is the point of Nils Christie's (1977) famous article 'Conflict as property'. In the penal system the conflict is between the offender and the state as a victim, and these are the two parties involved. In the sense that one can claim that there is a conflict involved between the animal and the buyer/trafficker, as one could legitimately claim in the case when humans are victims of trafficking (Sollund, 2012) and which makes sense since they have very conflicting interests, the animal will likely want to continue their life unharmed (Francione, 2008; Regan, 1983), while the offender's interest is to exploit them by making profit from enslaving them or by using them as a product. One could therefore claim that animals should be regarded as subjects with interest with a legal standing, as the victims they are in these cases.

This brings us to the debate of animal rights as discussed by Ted Benton (1998), and also to discussions by Gary Francione (2008) and

Taimie Bryant (2008), about whether animals can be persons. Francione (2008: 61) establishes that a person is one with moral interests, that the principle of equal treatment applies for a person, and that they shall not be regarded as an object. A person is one with moral rights. A condition for being regarded as a person with rights and not just a means in someone else's interest is that one is not property (Francione, 2008: 51). Because animals are regarded as property, when they 'belong' to the state like all confiscated wildlife in Norway belong to the Norwegian Environment Agency – hence the term 'poaching' – animals have no legal rights, even though animal welfare legislation exists (Benton, 1998; Bryant, 2008; Ellefsen, 2013). The animals are definitely not regarded as parts in a conflict; they are killed, they are regarded as none. In my material, they are instead objectified as evidence and illegal items and goods, even when they are alive. The amount of the fine depends on the economic value set for the animal(s). As shown, the punishment of offenders is consistently lenient. Not only are the animal victims regarded as insignificant but so are the breaches of law, and these two aspects of these cases are inextricably linked.

The wildlife trade from a green criminology perspective

I regard trafficking in animal and animal parts as a crime irrespective of their status as threatened and where this takes place, thus positioning my research within a non-anthropocentric, antispeciesist green criminology (White, 2013). No matter whether trade in a species is banned or regulated, whether these animals have been legally or illegally abducted from their habitats to be trafficked as pets, trophies, part of collections, mashed into powder as part of superstitious medicinal products or to give status to the owner, they are, or have been, individuals, and they are therefore victims of exploitation and abuse. Whether rights would provide a more secure platform for protection against direct and indirect animal abuse than, for example, the animal welfare legislation we have today is debatable (Benton, 1998). Yet within green criminology it is quite well established that animals are regarded in perspectives of justice, like species justice (e.g. Benton, 1998; Agnew, 1998; Beirne, 1999; White, 2011, 2013). Species justice acknowledges non-human animals' rightful place in their environment and their intrinsic right not to suffer from abuse, whether, one-to-one, institutionalized harm or harm arising from human actions that harm environments and climates (e.g. White 2012: 23). The growing concern regarding the survival of other animal species and their suffering caused by human exploitation is relatively recent in the context of the timespan of human history, and it has become especially pertinent with the rise of animal welfare movements. Those who recognize that animals, like humans, have intrinsic value and therefore should be protected often take the philosophy of Tom Regan as a point of departure, implying that animals also have interests in living a life that is free from abuse by humans. The constant killing of confiscated animals is due to the peculiar belief that animal welfare may be compatible with being killed, which fails to recognize that all living beings have an interest in life itself (Regan, 1983; Francione, 2014). The oddity of such a claim becomes evident if one does the thought experiment of putting a human in a non-human animal's position, thus recognizing that we also are animals and therefore in this regard have the same basic interests. However, in comparing the victims of human and animal trafficking, it is important to recognize that while we seek to prevent human trafficking in the interests of the victims, we seek simply to control animal trafficking so as to secure its endurance (Sollund, 2012).

In the case of wildlife trade, the actions whereby animals are abducted or killed are the first abuse they suffer. However, that abuse is facilitated by the overall doxic anthropomorphism which, for example, is the basis of the CITES convention. CITES can be accused of being a relatively modern convention while reflecting antiquated and outdated ideas about the human–animal relationship, or lack thereof.

Wildlife trafficking combines old-fashioned superstitions and practices (e.g. food practices and the medicinal value of certain animal species, such as rhino, Tiger, Black Bear and many reptile species (Alves et al., 2008; Santana, 2008; Richard, 2005; Minnaar, 2014)) with newer consumerist and materialist practices (e.g. the trafficking and trade for trophies and collections, such as serving wine with crushed rhino horn to show wealth and achieve status (Minnaar, 2014)). However, long before we humans became a threat to the entire world owing to our mere numbers and expansion, we left behind us large piles of bones of animal species that we drove to extinction (Hessen, 2013). Today the dramatic situation for animal species is clear from the decline of 39% in mammal, bird, reptile and amphibian species between 1970 and 2010 (WWF, 2014).

Conclusion

In green criminology it is acknowledged that a harm may be recognized as a crime even when it is not actually an infringement of law or regulation because what is criminalized will depend on who is in the power to define some acts as crime and others not, and this will vary with time and place (e.g. Lynch and Stretesky 2014; White 2007; Walters 2010). Following this the trafficking in free-born animals is a crime whether legal or illegal because whether infringement of law or not the act remains equally harmful for the victims involved. The need to discuss these cases of abuse, death and theriocide (Beirne, 2014) in a broad light to understand their motivation – such as the ideology, philosophy, culture and traditions behind both the regulations and the trafficking abuses – is urgent. In reality the regulations which should protect wildlife and the trafficking itself represent two sides of the same coin. The cases considered are interesting because of the total disrespect that they often show for individual animal lives, and this is shared by the CITES convention, by the offenders, by the control agencies through their lack of priority, through the treatment that these cases get in the judicial system, through the leniency with which they are punished, and not the least with the ultimate killing of the victims. For example, it appears that whether the cases are given priority in the judicial system which directs the punishment that the offenders are likely to get is very much dependent upon the discretion of both the police investigators and the prosecutors, and ultimately of the judges. Their priorities are politically determined: when the political target is war on drugs, then the wildlife trade will go undetected; and if occasionally detected it is not prioritized. The growing concern about the illegal wildlife trade which has been witnessed, for example, over the past couple of years, including the burning of stockpiles of ivory in France, Hong Kong, the USA and several African countries, seems so far not to have reached Norway. Creating awareness among consumers, whether they are in source or receiving countries, to increase empathy with the direct victims and the ecosystems to which they belong and with Mother Earth – *Pachamama* (Zaffaroni, 2013) – through care ethics (Adams and Donovan, 1996), rather than focusing on the economic value (and thereby economic losses) involved in the illegal wildlife trade, could be a move in the right direction. I therefore echo Brisman and South (2014: 34) in saying that green criminologists, rather than accepting constructions of crime as these are presented (and represented) in the mass media, must contribute to constructing crimes which have previously been undercommunicated as such, as the crimes they rightfully are, such as the trafficking in free-born animals, whether legal or illegal. This can be achieved through a focus on the direct victims of the trade more than on the indirect loss for humans because of the economic value that is attributed to the victims.

It is important to punish such acts because the law when enforced has a normative effect, and thus could serve to deter not least against the middlemen, organized groups and others who encourage the trade (for the typologies of the offenders, see Wyatt, 2013a). Norwegians and others who make a deliberate choice to take part in, or support, the illegal wildlife trade should receive a punishment which reflects the harm of the crime. However, punishing local indigenous peoples who abduct and kill non-human animals owing to a lack of alternative income, and who may have been encouraged to engage in such acts by middlemen, may be unjust from an environmental rights perspective. The middlemen, however, and those behind them should be punished in a way which deters them and others from getting involved in such practices. What measures should be taken against offenders must therefore be carefully assessed in each specific case. This requires increased priority within the judicial and control systems and at the political level globally, as this is a global problem.

Notes

1. Or in animals whose species have not been affected by human influence, even when bred by them, thus usually still referred to as 'wildlife' (Wyatt, 2013).
2. I dislike the anthropocentrism reflected in the term 'wildlife'. It is alienating, creates a gap between humans and non-humans, and implies that non-human animals are uncivilized and unpredictable in contrast to humans, who are therefore positioned above them. It is further a mass term which disguises the fact that free-born non-human animals are individuals (see Beirne, 2011). Likewise, the term 'animal' conceals the fact that humans also are animals, and that what is included in the term (by humans) is a large diversity of individuals of thousands of species, rather than one which are thereby contrasted with humans. For simplicity, however, and in the absence of good alternatives, both terms will still be used in this chapter, for which I apologize. I prefer to use the word 'kill' rather than 'euthanize', which is usually applied when animals are deliberately killed when they are not hunted, since I regard the latter as a euphemism that is used to conceal the true character of the act – that someone takes another innocent being's life.
3. http://www.cites.org/eng/disc/what.php.
4. The project has been running for four years as part of my research time as a professor at the University of Oslo. It has received funding from Dyrevelferdsfondet. The case study further forms part of the FP7 funded project EFFACE (http://efface.eu/).
5. 17 interviews: three with the eco-crime section of the police; one with a police officer, in addition to several informal conversations that were related to cases under investigation; one with two police officers (one police lawyer) regarding their investigation experience relating to environmental crimes;

two with the customs directorate; two with several customs officers at Oslo airport; two with veterinarians, one from the Norwegian Food Safety Authority (ironically responsible for animal welfare in Norway, indicating that animals are foremost regarded as food), working at Oslo airport where she takes care of animals that are stopped in trafficking; one with the chair of the CITES Standing Committee, the committee chair, who is also at the Norwegian Environment Agency in Norway, the institution that is responsible for CITES in Norway; and five with people who keep reptiles (illegally) and who have also trafficked reptiles to Norway.

6. The interviews in South America were carried out by David Rodrígez Goyes, my research assistant on the project. He also made the selection of informants under my guidance and recruited the interviewees. Interviews included Centro de Rehabilitación de Fauna Silvestre – the rehabilitation centre for wildlife, under the district secretary of the environment in Bogotá and Santa Fe. (I also did a SKYPE interview with people there, including Javier Cifuentes Álvarez) with Proaves, a conservationist NGO organization in Colombia, and Traffic in South America. Interviewees were from both NGOs and state agencies, and also included wildlife veterinarians and wildlife experts at universities. Data from Brazil included (group) interviews with DeFau – Secretaría de Medio Ambiente de Sao Paulo, the Chico Mendez Institute – part of the Ministry of Environmental Affairs, and Corredor das Onças and DeFau. A further interview was with Marcelo Robis Francisco Massaro, Capitán de la Environmental Military police of São Paolo. David also accumulated news articles, court rulings, police statistics and photographs of wildlife confiscations.

7. For a discussion of this priority, see the European Commission consultancy document prepared by Maher, Sollund and Fajardo, EFFACE, http://efface. eu/efface-contribution-commissions-consultation-wildlife-trafficking.

8. http://lovdata.no/.

9. http://lovdata.no/dokument/NL/lov/1981-05-29-38. This law was revised in 2009.

10. http://lovdata.no/dokument/NL/lov/2009-06-19-100.

11. http://lovdata.no/dokument/NL/lov/2009-06-19-97.

12. FOR-1976-11-20-3 Regulation of 20 November 1976 about the prohibition against the importation, commercialization or keeping in captivity of alien (exotic) animals (§1). Other regulations which apply are FOR-2004-02-20-464 about the welfare conditions for the import of live animals, http://lovdata.no/dokument/SF/forskrift/2004-02-20-464, http://lovdata.no/dokument/SF/forskrift/1976-11-20-3, and FOR-1991-07-02-507, regulation about prohibition of imports of animals and other contagious objects (sic), http://lovdata.no/dokument/SF/forskrift/1991-07-02-507.

13. http://lovdata.no/dokument/NL/lov/1997-06-06-32.

14. http://lovdata.no/dokument/SF/forskrift/2002-11-15-1276.

15. http://www.proaves.org/rubrique.php?id_rubrique=199.

16. This information was revealed by David Rodríguez Goyes.

17. I dislike the term 'pet' because it reduces animals to being someone for humans rather than for themselves, which gives them an instrumental rather than an intrinsic value (Sollund, 2008). Further in the case of reptiles, very often they cannot be handled because this will harm them, and

the purpose for keeping them is very often not physical and social proximity but rather to achieve reputation and status, as well as fascination.

18. It appears from a handwritten commentary by the customs directorate on the confiscation report that the same man was also stopped in 2000 with at least 27 birds, species unknown. Other data confirm that he is a recidivist.

19. I also did a Google search which supports this, as the man runs a firm registered in Norway which imports and breeds birds for sale to zoo shops and private people.

20. See, for example, 'Trafficking of 30 Endangered Jaco Parrots Prevented at Bulgarian Border'. Accessed at http://www.novinite.com/articles/163391/ Trafficking+of+30+Endangered+Jaco+Parrots+Prevented+at+Bulgarian+Border #sthash.9zv8NPiX.dpuf.

21. But whether wild caught or not, they were captives.

22. Traffic – for example. Accessed on 10 October 2014 at http://www.traffic.org/ home/2014/9/19/thailands-tortoise-seizures-significant-but-insufficient-fol .html.

23. In this case the Asian woman who smuggled it said that she wanted to test if there was money to be made from selling primate heads.

24. Accessed at http://www.osloby.no/nyheter/3000-kroner-dyrere-a-urinere-i-Askim-enn-i-Lillestrom-7244245.html.

25. With regard to Norwegian predators, see Sollund 2015 (not in print).

Bibliography

Adams, C. and Donovan, J. (1996) *Beyond Animal Rights. A Feminist Caring Ethic for the Treatment of Animals*. New York: Columbia University Press.

Agnew, R. (1998) 'The causes of animal abuse: A social-psychological analysis', *Theoretical Criminology*, 2(2): 177–209.

Alves, R.R. da N., a, da Silva Vieira and Santana, W.L. (2008) 'Reptiles used in traditional folk medicine: Conservation implications.' *Biodiversity and Conservation*, 17: 2037–2049.

Alves, Rómulo Romeu da Nóbrega, Washington, Luiz da Silva Veira and Gindomar Gomes Santana (2008) 'Reptiles used in traditional folk medicine: Conservation implications', *Biodiversity and Conservation*, 17: 2037–2049.

Benton, T. (1998) 'Rights and justice on a shared planet: More rights or new relations?' *Theoretical Criminology*, 2: 149.

Beirne, P. (1999) 'For a non speciesist criminology: Animal abuse as an object of study', *Criminology*, 37(1): 117–149.

Beirne, P. (2011) 'Animal abuse and criminology. Introduction to a special issue', *Crime, Law and Social Change*, 55: 349–357.

Beirne, P. (2014) 'Theriocide: Naming animal killing', *International Journal for Crime, Justice and Social Democracy*, 3(2): 46–66.

Brisman, A. and South, N. (2014) *Green Cultural Criminology. Constructions of Environmental Consumerism and Resistance Too Ecocide*. London: Routledge.

Bryant, T. (2008) 'Sacrificing the sacrifice of animals: Legal personhood for animals, the status of animals as property. And the presumed primacy of humans'. *Rutgers Law Journal*, 39: 247–330.

Burgess. E.A., Stoner, S.S. and Foley K.E. (2014) *Brought to Bear: An Analysis of Seizures Across Asia (2000–2011)*. Selangor, Malaysia: Traffic, Petaling Jaya.

Castañaga, M., Macleod, R., Hennessey, B., Núñez, J. U., Puse, E., Arrascue, A., Hoyos, J., Chambi, W. M., Vasquez, J. and Engblom, G. (2011) 'A study of parrot trade in Peru and the potential importance of international trade for threatened species', *Bird Conservation International*, 11: 76–85.

Christie, N. (1977) 'Conflict as property', *British Journal of Criminology*, 17(1): 1–15.

CITES convention. Accessed on 14 June 2011 at http://www.cites.org/eng/disc/text.shtml.

Drews, C. (2002) Attitudes, knowledge and wild animals as pets in Costa Rica. *Anthrozoos: A Multidisciplinary Journal of the Interaction of Peoples and Animals*, 15(2): 119–133.

Du Plessis, M.A. (2000) 'CITES and the causes of extinction'. In Hutton, J. and Dickson, B. (eds) *Endangered Species Threatened Convention*. London: Earthscan.

Eniang, E.A., Akpan, C.E. and Eniang, M.E. (2008) *A Survey of African Grey Parrots (Psittacus Erithacus) Trade and Trafficking in Ekonganaku Area of Ikpan Forest Block, Nigeria*.

Ellefsen, R. (2013) *Med rett til å pine*. Om bruk og beskyttelse av dyr. [Entitled to torture. About use and protection of animals]. Oslo: Fritt forlag.

Ellis, R. (2005) *Tiger Bone & Rhino Horn. The Destruction of Wildlife for Traditional Chinese Medicine*, Washington: Islands Press. Shearwater Books.

European Commission (2014) Communication from the commission to the council and the European Parliament on the EU Approach against Wildlife Trafficking. Accessed on 20 October 2014 at http://ec.europa.eu/environment/cites/pdf/communication_en.pdf.

Foley, J.A. (2013) 'Poor Regulation, Thriving Pet Trade Threatens Monitor Lizard Populations in SE Asia', *Nature World News*. Accessed on 26 September 2014 at http://www.natureworldnews.com/articles/2241/20130603/poor-regulation-thriving-pet-trade-theatens-monitor-lizard-populations-se.htm.

Francione, G. (2008) *Animals as Persons. Essays on the Abolition of Animal Exploitation*. New York: Columbia University Press.

Francione, G. (2014) 'Animal welfare and the moral value of nonhuman animals'. In Cederholm, A.E., Björck, A., Jennbert, K. and Lönngren, A.-S. (eds) *Exploring the Animal Turn*. Lund: The Pupendorf Institute.

Gonzales, J.A. (2003) 'Harvesting, local trade, and conservation of parrots in the Northeastern Peruvian Amazon', *Biological Conservation*, 114(2003): 437–446.

Guzman, J.C.C., Sanchez, S.M.E., Grosselet, M. and Gamez, J.S. (2007). The Illegal Parrot Trade in Mexico Defenders of wildlife. Accessed on 25 February 2013.

Hagstedt, J. and Korsell, L. (2012) 'Unlawful hunting of large carnivores in Sweden.' In Ellefsen, R., Sollund R. and Larsen, G. (eds) *Eco-global Crimes. Contemporary Problems and Future Challenges*. London: Ashgate.

Herrera, M. and Hennessey, B. (2007) 'Quantifying the illegal trade in Santa Cruz de la Sierra, Bolivia, with emphasis on threatened species', *Bird Conservation International*, 17: 295–300.

Hessen, D. (2013) 'Hvor unikt er mennesket?' [How unique is man?]. In Sollund, R., Tønnessen, M. and Larsen, G. (eds) *Hvem er villest i landet her?* [Who is the Wildest in This Country?]. Oslo: Spartacus SAP.

Hutton, J. and Webb, G. (2005) 'Crocodiles: Legal trade snaps back', In Oldfield (ed.) *The Trade in Wildlife. Regulation for Conservation*. London: Earthscan, pp. 11–120.

IFAW (International fund for animal welfare) (2008) *Killing with keystrokes. An investigation of the Illegal wildlife Trade on the World Wide Web*. Accessed on 14 June 2011 at http://www.ifaw.org/Publications/Program_Publications/ Wildlife_Trade/Campaign_Scientific_Publications/assets_upload_file64_12456. pdf.

Lee, J. (1996) 'Poachers, tigers and bears . . . Oh my! Asia's illegal wildlife trade', *Northwestern Journal of International Law and Business*, 498(16): 495–515.

Lemieux, A.M. and Clarke, R. (2009) 'The international ban on ivory sales and its effects on elephant poaching in Africa', *British Journal of Criminology*, 49: 451–471.

Lynch, M. and Stretesky, P.B. (2014) *Exploring Green Criminology. Toward a Green, Criminological Revolution*. Farnham: Ashgate.

Metz, S. (2007). Rehabilitation of Indonesian parrots from the illegal wild bird trade: Early experience on Seram Island, Indonesia. *Proceedings. Association of Avian veterinarians, Australian Committee*. Accessed 29 June 2010 at http://www. indonesian-parrot-project. org/pdf_files/AAV_07_6-29-07. PDF.

Miljøkrim (2013) 'Finlands mest omfattende eggsak', *Miljøkrim*, 1 April 2013.

Minnaar, A. (2014) 'The poaching of rhino in South Africa: A conservation, organised and economic crime?'. In Sorvatzioti, D., Antonopolous, G., Papanicoulaou, G. and Sollund, R. (eds) *Critical Views on Crime, Policy and Social Control*. Nicosia: University of Nicosia Press.

Nurse, A. (2013) *Animal Harm. Perspectives on Why People Harm and Kill Animals*. Aldershot: Ashgate.

Regan, T. (1983) *The Case for Animal Rights*. Berkeley University Press.

Reid, W.R. (1992) 'How many species will there be? Chapter 3'. In Whitmore, T.C. and Sayer, J.A. (eds) *Tropical Deforestation and Species Extinction*. New York: Chapman and Hall.

Rosen, G. E. and Smith, K. (2010) 'Summarizing the evidence on the international trade n illegal wildlife', *Ecohealth*, 7(1): 24–32.

Runhovde, S. (forthcoming) 'A pleasant surprise? Policy, discretion and accidental discoveries in enforcement of illegal wildlife trade at the Norwegian border'.

Santana (2008) 'Reptiles used in traditional folk medicine: Conservation implications', *Biodiversity Conservation*, 17: 2037–2049.

Sheperd, C. Stengel, C.J. and Nijman, V. (2012) *The Export and Reexport of CITES Listed Birds from the Solomon Islands*. A Traffic Southeast Asia Report. Rufford: Traffic.

Svae-Grotli, I. (2014) – En forbrytelse å la alvorlig miljøkriminalitet forbli forseelser [–A crime to allow serious environmental crime remain misdemeanours] In *Miljøkrim* 1: 30–36.

Sollund, R. (2011) 'Expressions of speciesism: The effects of keeping companion animals on animal abuse, animal trafficking and species decline', *Crime, Law and Social Change*, 5: 437–451.

Sollund, R. (2012) 'Speciesism as doxic practice'. In Ellefsen, R., Sollund R. and Larsen, G. (eds) *Eco-global Crimes. Contemporary Problems and Future Challenges*. London: Ashgate.

Sollund, R. (2013) 'Animal trafficking and trade: Abuse and species injustice'. In Walters, R., Westerhuis, D. and Wyatt, T. (eds) *Emerging Issues in Green Criminology. Exploring Power, Justice and Harm.* Basingstoke: Palgrave.

Sollund, R. (2015) 'With or without a license to kill: Human-predator conflicts and theriocide in Norway'. In Brisman, A., South, N. and White, R. (eds) *Environmental Crime and Social Conflict: Contemporary and Emerging Issues.* Aldershot: Ashgate.

Stoett, P. (2002) 'The international regulation of trade in wildlife: Institutional and normative considerations', *International Environmental Agreements. Politics, Law and Economics,* 2(2): 195–210.

South, N. and Wyatt, T. (2011) 'Comparing illicit trades in wildlife and drugs', *An Exploratory Study Deviant Behavior,* 32(6): 538–561.

Svae-Grotlie, I. (2014) 'En forbrytelse à la alvorlig miljøkriminalitet forbli forseelser', *Miljøkrim,* 1(17): 30–37.

Tidemann, S. and Gosler, A. (eds) (2009) *Ethno–Ornithology: Birds, Indigenous Peoples, Culture and Society.* New York: Earthscan.

Vecchiatto, P. (2014) 'WWF launch drive to cut demand for rhino horn'. Accessed on 26 September 2014 at http://www.bdlive.co.za/national/science/2014/09/23/wwf-launch-drive-to-cut-demand-for-rhino-horn.

Walters, R. (2010) 'Crime is in the air. The Politics and Regulation of Pollution in the UK.' London.

Warchol, G. (2007) 'The transnational illegal wildlife trade', *Criminal Justice Studies: A Critical Journal of Crime, Law and Society,* 17(1): 57–73.

Warchol, G.L., Zupan, L.L. and Clarke, W. (2003) 'Transnational criminality: An analysis of the illegal wildlife market in Southern Africa', *International Criminal Justice Review,* 13: 1–26.

Wasser, S.K., Clark, B. and Laurie, C. (2009) 'The ivory trail', *Scientific American,* 301 (June 2009): 68–76.

Weston, M.K. and Memon, M.A. (2009) 'The illegal parrot trade in Latin America and its consequences to parrot nutrition, health and conservation', *Bird Populations,* 9:76–83.

White, R. (2007). "Green criminology and the pursuit of social and ecological justice" In Beirne, P. and South, N. (eds) *Issues in Green Criminology. Confronting Harms against Environments, Humanity and Other Animals.* Devon: Willan.

White, R. (2011) *Transnational, Environmental Crime. Toward an Eco-global Criminology.* London: Routledge.

White, R. (2012) 'The foundations of eco-global criminology'. In Ellefsen, R., Sollund, R. and Larsen, G. (eds) *Eco-global Crimes. Contemporary Problems and Future Challenges.* Farnham: Ashgate.

White, R. (2013) 'The conceptual contours of green criminology'. In Walters, R., Westerhuis, D. and Wyatt, T. (eds) *Emerging Issues in Green Criminology. Exploring Power, Justice and Harm.* Basingstoke: Palgrave.

Wilgen, N.J., van Wilson, J.R.U., Wintle, B.A. and Richardson, D.M. (2010) 'Alien invaders and reptile traders: What drives the liv animal trade in South Africa?' *Animal Conservation,* 13(1): 24–32.

Wright, Timothy et al. (2001) 'Nest poaching in neotropical parrots', *Conservation Biology,* 15(3): 710–720.

Wyatt, T. (2013a) *Wildlife Trafficking: A Deconstruction of the Crime, the Victims and the Offenders.* Basingstoke: Palgrave.

Wyatt, T. (2013b) 'The local context of international wildlife trafficking: The Heathrow Animal Reception Centre'. In Walters, R., Westerhuis, D. and Wyatt, T. (eds) *Emerging Issues in Green Criminology. Exploring Power, Justice and Harm.* Basingstoke: Palgrave.

WWF 'More Tigers in American Backyards than in the Wild'. Accessed on 24 September 2014 at http://www.worldwildlife.org/stories/more-tigers-in-american-backyards-than-in-the-wild.

WWF (2014a) Overview. Accessed on 25 September at http://www.worldwildlife.org/threats/deforestation.

WWF (2014b) *Living Planet Report 2014. Species and Spaces, People and Places.* Accessed on 9 March 2015 at https://www.wwf.or.jp/activities/lib/lpr/WWF_LPR_2014.pdf.

Zaffaroni, E.R. (2013) *La Pachamama y el humano.* Ediciones Madres de Plaza de Mayo: Ediciones Colihue.

Zhang, L., Ning, H.N. and Sun, S. (2008) 'Wildlife trade, consumption and conservation awareness in southwest China', *Biodivers Conservation,* 17: 1493–1516.

Zimmerman, M.E. (2003) 'The black market for wildlife: Combating transnational organized crime in the illegal wildlife trade', *Vanderbilt Journal of Transnational Law IVOL,* 36: 1657.

9

Denying the Harms of Animal Abductions for Biomedical Research

David Rodríguez Goyes

> The truth is that I was not able to sleep. I was not able to sleep knowing that three blocks from my house, illegal experiments on owl monkeys were being conducted.
>
> (Ángela María Maldonado, personal interview)

Introduction

When people are confronted with information that is deemed to be so disturbing that it is difficult to remain passive, acknowledgment responses are elicited. By becoming committed to the issue, an act of acknowledgment provides a psychologically and morally appropriate response to what the person knows. However, if a harmful situation exists but its occurrence, its meaning or its implications are effectively denied, then actions against it are avoided and the possibility is created for the situation to persist. Consequently, the practices in question are likely to keep on occurring if they are not recognized as harmful or if their harms are denied (Cohen, 2001).

Originally the concept of denial was regarded as a psychological defence mechanism that is used to unconsciously deal with the guilt or anxiety that is produced by unbearable information (Cohen, 2001). Through it, troubling information is rejected. In the sociology of denial developed by Stanley Cohen, the concept is expanded to cover all of the situations in which, consciously or unconsciously, the agent tries to shut down the threatening information that indicates that something is deeply wrong. The agent might shut down such information by denying events, giving the event an alternative interpretation or justifying their happening, and thus they can avoid reacting to it. The motivation behind such mechanisms, whether conscious or unconscious, is the desire to provide for ourselves a state of wellbeing that would be threatened by the acknowledgment of troubling information.

Denial allows someone to commit an act and meanwhile sway observers into not taking action. The denials resorted to by them of a crime/harm are conducive to allowing the atrocity to take place. For example, acid-attackers[1] deny the blameworthiness of their acts by assuming that women are objects for them to use and destroy at will, so even when recognizing the harms that they cause they consider them to be justified (Rodríguez Goyes, 2013a). This denial makes it easier to commit the assault because it allows the perpetrator to avoid the psychological disturbance that would normally follow. When they accept the officially instituted rules of behaviour, this process could be understood as an application of a set of 'techniques of neutralization' through which, by justifying the act, internal and external disapproval are deflected (Sykes and Matza, 1957). The denials developed by bystanders are equally decisive in the production and persistence of harm. A large percentage of environmental harms do not invoke a reaction from people who are aware of these harms. It is not uncommon that this lack of reaction is produced by the conviction that it is impossible that something similar will happen to them because they assume that the victims are different from themselves and because of the social distance that they have to them on that account[2] (see, e.g., Agnew, 1998).

Denials can be personal, official or cultural. At the personal level, they are individually sustained. Official denials are public, collective and highly organized distortions and deceptions that are sustained by official authorities through the resources of modern states. In it, both the facts and the meanings are denied, and if recognized, the harms are explained away as isolated events or beyond government control (Cohen, 2001). Cultural denials are unwritten arguments that are created by the majority within a society about what can be acknowledged and what will be denied in order to maintain their privileges (Cohen, 2001).

In 2009, during the 37th Annual Conference of the European Group for the Study of Deviance and Social Control, Stanley Cohen presented a paper entitled 'Panic or denial: On whether to take crime seriously' (2013). The basic questions to be addressed were when an event is bad enough to deserve an action in defence of what is being harmed (acknowledgement), and when a complaint is not as important as it has been portrayed, so that it is possible to refer to it as an exaggeration or a case of moral panic (Cohen, 2011). To answer these questions, a basic measurement scale or a moral basis of 'deservedness of acknowledgment' is required. Providing a yardstick to measure how much an event deserves acknowledgment is difficult. To begin with, as constructionists (Wallner, 1994) and interactionists (Goffman, 1997; Becker, 2010) have shown, there is not a single truth but many, all of which are socially constructed, making it impossible to measure the accuracy of one interpretation of reality against another. Moreover, the idea of the existence of an ontological morality has been claimed as wrong (see, e.g., Nietzsche, 2005).

From a green criminological perspective – a field of study that is concerned with 'transgressions against ecosystems, humans and non-humans' (White, 2012) – it is imperative to attend to such fundamental questions regarding what issues merit recognition as being harmful or problematic (White, 2013). The only possible starting point for preventing harmful practices is to recognize them as such and then to confront possible damaging denials.

In view of the above, in this chapter I present the Colombian Patarroyo case, to be used as a useful example (Stake, 2005) to advance the search for adequate guidelines for contextually determining when a situation deserves acknowledgment as being harmful and illegitimate. The Patarroyo case, named after the doctor who led the research, revolves around experimentation on animals[3] of the *Aotus* genus[4] that were directly abducted from the forest in order to develop a malaria[5] vaccine. It shows that animal experimentation in biomedical research is an important source of harm inflicted on animals, environments and humans. However, these abusive practices go largely unquestioned (Sollund, 2012a) because of the backing provided by the modern idea of progress, through which the myriad of harms produced are denied, and their opponents are labelled as moralistic or blinded by environmental sentiments. As such, the proposed task in this chapter is to examine the moral bases that are used to justify both the denial and the acknowledgment reactions that are at play in this particular case.

A warning about the terminology

As a long tradition in critical criminology has suggested, we access reality through a particular set of discourses (see, e.g., Hulsman, 2003; Hulsman and Bernat de Celis, 1984; Hall, 1997). Consequently, concepts that are used to designate non-human animals as objects or goods are conducive to harmful practice by sustaining an abusive relationship between human beings and all other species (Sollund, 2013a). While I am conscious of the problematic use of speciesist language, in order to stay as close as possible to the legal and administrative terminology that was originally used in this case, when reporting the facts I use terminology such as 'resources', 'capture of specimens', 'extractions' or 'invasive species'. Such language is anthropocentric and problematic because it reduces the life and intrinsic value of the species concerned. The correct terminology to be used when referring to any species within the environment, in my view, is one that upholds the intrinsic worth that each of them possess. An important example of more adequate language is the word 'abductions', which is used in the title of this chapter. It was first used concerning animals by Sollund (2011) and it refers to the violent act of forcing any sentient being to go with you against their will. As such it recognizes the intrinsic right of freedom to all beings in the environment – whether human or not.

The Patarroyo case[6]

In 1987, pathologist Manuel Elkin Patarroyo announced the development of the world's first synthetic malaria vaccine (SPf66). For his contribution to humanity he was given the Prince of Asturias Award. In an allegedly magnanimous act, he donated the intellectual property of his vaccine to the World Health Organization (WHO). Not everything that followed was as commendable. Years later the SPf66 vaccine was described as a dud. While Patarroyo rated it with 30% efficacy, others were harsher: 'no protection by SPf66 vaccines against P. falciparum[7] in Africa...a modest reduction in attacks of P. falciparum malaria following vaccination with SPf66 in South America' (Graves and Hellen, 2006).

Patarroyo's scientific quest dates back to 1984, when INDERENA[8] granted him a *Study of Biological Diversity Permit*, allowing him to capture 200 individuals of the *Aotus vociferans* species from the Colombian side of the Amazon basin over a two-year period. The purpose of abducting the monkeys was to use them as objects in the experimentation that would lead to the development of a synthetic vaccine against the malaria disease. New permits were granted to the Colombian Institute of Immunology (FIDIC), a private foundation whose head is Manuel Elkin Patarroyo, this time by Corpoamazonía, in the years of 1999, 2002, 2006, 2010, totalling the allowance to collect 8,200 owl monkeys over a discontinuous period of 15 years. In 1999, notwithstanding his initial failure and counting on the political backing of Spanish Royalty (Desowitz, 2002), Patarroyo announced to *The Guardian* that he was already testing a new vaccine – a 100% efficient malaria vaccine.

'He has used 11,500 monkeys to test his 100 per cent vaccine and his work is being reviewed', reported the newspaper (Brown, 1999). Some 15 years later, no 100% vaccine has been released but thousands of monkeys have been added to the 11,500 that have reportedly abducted up to 1999.

In order to 'supply' the experimentation facility there are two methods for collecting monkeys from a tree where a nest is located. The first is an especially difficult one and requires that various collectors stealthily ascend through the lateral branches of the tree, and once at its top they block all possible exits so that when individuals try to get out they can abduct them. This technique is rarely used as only a small number of skilful indigenous people have the necessary skills. The other, more common, method is to cut all of the trees in a 300–500 m² circle around the tree containing the nest, leaving a single path of trees. A fishing net is placed at the end of the path. Loud noises are made so that the scared monkeys flee from the nest and fall into the net. An estimated 65,000 trees are consequently lost annually with the sole purpose of providing the FIDIC with monkeys (Maldonado and Peck, 2014), creating both a direct and an indirect negative impact on biodiversity.

To gather the required monkeys for its research the FIDIC paid locals US$ 40 per owl monkey captured. Hunting owl monkeys is not a traditional practice as the animals are regarded as small, meatless and distasteful. Moreover, hunting monkeys other than for survival is considered to be a breach of ancient rules in traditional Amazonian communities. However, after three decades of this activity, around 133 collectors from 19 indigenous Amazonian communities derive their income from selling monkeys for research (Maldonado, 2014). To disregard tradition and to impose economic activities that leave only transient monetary profit could be regarded as a continuation of the chain of exploitation of impoverished communities. Moreover, having the expectation of receiving money to aid their economies, collectors are always in search of monkeys – the more they collect the more money they receive. Thus, allegedly, the number of captured monkeys is far more than the number allowed. For example, between the months of March and May 2012, 912 monkeys were registered for admission into the FIDIC facilities even when the hunt permit allowed at most 800 owl monkeys to be captured per year (Maldonado, 2014).

When returning the animals from the FIDIC laboratories back to the forest there has been no compliance with IUCN guidelines (IUCN/SSC, 2013) – they are released in places that are unsuitable for them and in numbers 40 times as big as the average group (3–5), and screening and rehabilitation processes are non-existent or insufficient (Maldonado and Peck, 2014). Moreover, Patarroyo is experimenting in Colombia with a non-endogenous malaria strain. If the pathogens that are injected into the monkeys reappear, and from them vectors infect other animals or human populations, these practices could create a biosecurity problem (Paz-y-Miño, 1997).

Furthermore, given that 'for the granting of these permits no scientific study was done, thus, there is no certainty if the quota of individuals collected puts a risk to the existence of the natural resource [*Aotus vociferans*]' (Acción Popular, 2013), the result, as professor Marta Bueno states, is that the species is threatened with extinction (personal interview, 24 June 2014). This is not mere rhetoric when taking into consideration that, although study permits were granted by the Colombian authority, and thus authorization was granted only to collect individuals of *Aotus vociferans*, which is listed in CITES II as being present in Colombia, owing to the economic incentives involved, inhabitants of Brazil and Peru have also been seen hunting and selling monkeys. As a result, members of another species – *Aotus nancymaae* – which is not recognized by CITES as Colombian, were accepted by FIDIC (Acción Popular, 2013; Maldonado and Peck, 2014). These events represent a breach of CITES[9] to which Colombia has signed up, and are moreover defined as a crime according to Colombian criminal law[10]. Hence, after extracting more than 15,000 monkeys over three decades, to this must be added the fact that *Aotus vociferans* now has to fight over territory with *Aotus nancymaae*, so the likely outcome is a rapid decline in their populations.

With regard to the monkeys' treatment at the research site, Patarroyo stated the following at the Colombian State Council chambers: 'They are requesting us to give the animals a good treatment, and we have always done it.'[11] Evidence indicates otherwise. Monkeys are subject to a myriad of mistreatments throughout the entire process: through the act of abduction they are frightened and displaced from their nests; even when they live in small families composed of a male, female and their offspring, they are separated; in spite of the knowledge that both species are nocturnal, they are kept in laboratories that are fully lit during the day; they are held in cages; their spleens are extracted; and they are injected and purposely infected with malaria. Nestor Roncancio from the Institute for Scientific Research of the Amazon reported[12] that monkeys released from the FIDIC facilities have lost a fifth of their original body weight. Finally, as described by Maldonado and Peck (2014), 'local people reported the presence of carcasses of owl monkeys close to their crops. They described animals having a tattooed number on their legs, which corresponds to the code given by the laboratory during the experimentation.'

The Patarroyo case thus encompasses a whole set of practices that have been engaged in from as early as the 1980s in the search for a malaria vaccine. The harms in question are rooted in cultural structures, and political and economic relations of Western societies (Stretesky et al., 2014).

Malaria research should be examined in view of the structural context in which it takes place and takes shape. 'Malaria is an economic, social and cultural issue' (Carlos Agudelo in Cariboni and Viera, 2007), not only because of the consequences of the disease but also because of its causes and, of interest to us, the dynamics and effects of the search for a vaccine against it. In 1956, Colombia created the National Malaria Eradication Service (SNEM) (Hernández Álvarez and Obregón Torres, 2002). The SNEM's objective was to prevent malaria transmission by providing mosquito screens to communities and teaching them how and when to use them. Notwithstanding its success, early in the 1990s, when a supposedly effective malaria vaccine was found, the money that had previously been invested in preventive malaria campaigns was dramatically reduced and only a fraction of it was reinvested in vaccinations with SfP66. The results were far from satisfactory. According to the WHO, reported cases of malaria in Colombia rose from 99,489 in 1990 to 185,455 in 1999, almost doubling its rate of incidence (reported cases/total population). The highest rates were, and are, present in the most impoverished locations (WHO, 2013a). The false promise of an effective vaccine, added to policies that favour healthcare privatization, resulted in the closure of the SNEM in 1993 and with it the abandonment of those communities that were most in need.

Not only did the false announcement further exacerbate the effects of existing disparities in the distribution of access to preventive healthcare, educational campaigns and mosquito screens, but also the research methods used by the FIDIC contributed to sustaining structural inequalities and exploitative logics. The FIDIC introduced a new way of dependence in a region that was historically and systematically exploited[13]. As Robert Desowitz (1991, 1997) points out, a curious particularity of Western societies is that viruses are considered exclusively in terms of their biomedical nature, thus ignoring their other dimensions. For example, when trying to sum up the main issue about the malaria disease, Dr Manuel Elkin Patarroyo said[14] that '17 million people have died because of illnesses; if vaccines against them existed, they would still be alive'. The reality is, however, that viruses are intertwined with sociopolitical, economic and environmental contexts. Believing that the mere existence of a vaccine is an immediate solution to illnesses means, among other things, disregarding the fact that the poorest and most endangered part of the population has greater difficulty in accessing basic medical care than the rest primarily due to healthcare privatization (see, e.g., Hernández Álvarez and Torres-Tovar, 2010).

The main factors in effective malaria prevention are access to screened houses, preventive education and adequate healthcare. As biologist Marta Bueno explains, 'if you break the chain of the disease, if you stop the vector that stings a human and then stings another, then the disease is prevented' (personal interview, 24 June 2014). Proof of this is that whereas we see that malaria can indeed be effectively dealt with, thus far it has only been fully eradicated in the more affluent countries (e.g. Italy and the USA were the first to eradicate it), whereas the problem is still severely affecting the world's most marginalized communities (Desowitz, 1991, 2002). The fact that a large part of the world's population still suffers from this disease is the outcome of a set of political and economic relations and decision-making processes rather than an 'unfortunate' situation. Improving access to healthcare, preventive health education and screened housing in impoverished communities would mean a structural and distributive shift that is incompatible with the privatization of communal goods and services that is brought about by current dynamics of neoliberal capitalism.

Instead of applying these real preventative measures, a multibillion dollar industry has developed wherein antimalarial drugs are being manufactured and supplied around the world, with pharmaceutical companies profiting from selling their patented products, most of which were created through biopiracy processes (Desowitz, 2002).[15] Rather, a reactionary process has been put in motion: the search for a malaria vaccine. From as early as 1919, when efforts to create a vaccine began, up until now, research into the development of a malaria vaccine has been a path filled with disappointments. As Sanjeev Krishna puts it, 'The landscape of malaria vaccine development is littered with carcasses, with vaccines dying left, right and centre' (quoted in Mundasad, 2014).[16]

Denial

In modernity the road of human progress is defined and hoarded by positivist science, technique and technology. Thus positivist science interferes in all realms of life by imposing its rules and truths (Hoyos Vásquez, 2006). From a modern point of view, progress is seen as a unitary process that is arrogated by positivist science and its technologies, which through the absorbent appropriation of all realms of life is supposed to lead towards human completion (Vattimo, 1990). As reflected, for example, in the logic of CITES, non-humans are converted into resources for human fulfilment (Sollund, 2011), appropriating and controlling them (Mol, 2013). In CITES, concepts of natural resources and sustainable development appear. Natural resources refer to nature that is seen as commodities and that is valued according to human interests. Sustainable development notions refer to the methods that have been designed to ensure the endless human exploitation of nature (White, 2013). As such, non-human animals are transformed into commodities, therefore denying that they are equal to humans in trying to avoid pain, having an interest in not being harmed and wanting to carry on with their lives (Sollund, 2013b, 2014). Clara Henao, a biologist with extensive experience of working with Amazonian communities and who in 2003 was hired to work at the Corporation for the Sustainable Development of the Amazonian Region (Corpoamazonía), realized that in the Study of Biodiversity Permit the actual subject relating to the permit is the least important part of the procedure. For example, in the requirements regarding the management of disposals, the cleanliness of the facility and so on, none of the stipulations refer to the monkeys' actual wellbeing (personal interview, 27 May 2014).

Modern conceptions of the environment being a natural resource to exploit are reflected in Colombian law. Article 80 of the Colombian Political Constitution states: 'The State will plan the use and management of natural resources in order to warrant sustainable development.' Likewise, Decree 309 of 2000, which allows experimentation on wildlife when a permit is granted, is based on the consideration that the state shall warrant research and science and thus facilitate scientific practices (del-Medio-Ambiente, 2000). Of significance is the fact that there is an established quota of monkeys that are legally allowed to be abducted, because it implies that abducting them is not considered to be problematic as long as it does not exceed a certain threshold and reflects the anthropocentric logic that informs Colombian law, in which we only care about them because we need to conserve a portion of the population now so that we are able to use them in the future.

During her work, Clara Henao was assigned to follow the cases in which permits for the study of biological diversity were granted, to ensure that their holders were obeying the legal requirements that were attached to the authorizations. In 2004 she received the expedient of the FIDIC. She describes her first visit to the FIDIC facilities as follows:

I almost died. It was tragic. Any person, even the more insensitive one, would have been terrified. It was horrible: animals portraying sadness were caged as if in prison. There were many cages with monkeys inside; there were as well surgery rooms were monkeys' spleen was extracted. The most incredible part to me was that the girl in charge of taking care of them told me 'here the monkeys are in a five star hotel, we feed them with compote and we bathe them with detergent, they cannot complain'.[17]

Also in 2003, Ángela Maldonado[18] was in the Amazon basin as part of her doctoral studies in primatology. A short time before leaving the rainforest, while navigating the Amazon River, she caught a glimpse of what appeared to be a sack filled with monkeys being loaded onto an indigenous raft. Out of curiosity she asked the aboriginals about the contents of the sack. With a knowing smile they replied that they were carrying the monkeys to be delivered to the FIDIC, where they are used as subjects for malaria vaccine trials. Their answer provoked her interest, questioning the legality of such a practice. She immediately went to the Ministry of the Environment to denounce what was happening in the Amazon basin. The woman in charge at the ministry replied that it might have been a one-off event and that they would react only if she provided further proof.

Maldonado was not the first to encounter government denials. Others before her found even harsher reactions. An informant, whose name will be maintained in confidentiality for security reasons, told me how, after denouncing what was happening around the FIDIC facilities, 'a colonel of the military forces called, warning that doctor Patarroyo should be left alone, and that anyone in his way would disappear from the Amazon'. Less harsh but equally dramatic was the case of Clara Henao, who, with the help of the police, seized the monkeys from the FIDIC facilities. Immediately afterward the governor of the Amazon Department, the general director of Corpoamazonía and even the Colombian president were making calls in order to have the monkeys returned to the FIDIC. Henao was transferred and further access to the FIDIC case file was denied. Finally, a report made by Marta Bueno and Claudia Brieva (2007), warning about the possible extinction of *Aotus vociferans* and recommending that hunting activities involving the monkey were halted, disappeared from the FIDIC file case held at Corpoamazonía headquarters.

As a last resort, Maldonado, along with Gabriel Vanegas, lodged a popular lawsuit, arguing that the activities developed by the FIDIC were violating the collective rights to have a secure and healthy environment, to have a rational use and management of natural resources and a balanced environment. In the reply to this lawsuit, Colombian government authorities kept denying the harm (Acción Popular, 2013). The Ministry of the Environment replied that it was not the competent authority to manage the case, because law stipulates that Corpoamazonía is the body in charge. Corpoamazonía replied that all permits were granted according to law, and that it had exercised oversight so as to ensure that all procedures were legal. Similarly, the FIDIC replied that it had always acted according to the law.

In the ruling (Acción Popular, 2013) of the popular interest lawsuit filed by Maldonado, the Colombian State Council considered that the FIDIC practices constituted a breach of the collective rights to a safe and balanced environment and the rational use of natural resources. Consequently, it annulled the Study of Biodiversity Permits. No reference was made to the suffering of the animals, only to the inexistence of a study to determine how many monkeys could be extracted without affecting the ecosystem. The ruling left the door open for new permits: 'In the event to grant new permits to the FIDIC, all legal requirements shall be fulfilled' (Acción Popular, 2013).

This ruling (as well as Colombian environmental law) is based on the modern ideology of progress, in which nature is seen as a resource and positivism is the ultimate path to reach human wellbeing. To an extent, this is also reflected in some of the discourses and practices of the environmental justice perspective – a movement which began in 1982 within the movement against environmental racism in the USA (Schlosberg, 2007) and was later drawn upon by green criminology (Lynch, 2006) – whose propositions depart from an anthropocentric ideology that asserts that the environment must only be preserved because it is necessary for the wellbeing of present and future human beings (White, 2013). It is intrinsically a perspective that assumes that environmental rights are a prolongation of human rights, something that in theory should not exclude justice for the environment and the species within it (Sollund, 2013c). However, in its most common application it turns out to be applied in an anthropocentric fashion. As such, ultimately, human interests predominate in the environmental justice perspective, which means that these take priority over the interests of the environment and non-human animal species, something that supposedly would secure the most basic needs of the human being. However, a dualistic logic is implied in the environmental justice perspective, which carries the risk that some groups within the human population or non-human environment are given preferential treatment over others (Sollund, 2012b), that in this line of reasoning are met with a certain connotation of inferiority, backing their abusive treatment as 'resources' (see, e.g., Svärd, 2012; Rodríguez Goyes, 2013a). While such a perspective tries to defend the right for us and future generations to use a healthy environment, it also backs the ideology that contributes to the destruction of the environment and the species within it.[19]

Even while the ruling was apparently the acknowledgment of a wrongful situation, it is in itself a stronger official and cultural denial of the intrinsic value of all individuals within the environment. The Colombian State Council stated: 'It is not possible to accord full dignity to non-human animals and vegetal species, as this would mean that the human kind would not be able to use them for its survival' (Acción Popular, 2013). As Sollund warned in relation to this same ruling, 'No one can have dignity without rights. Being deprived of rights is per se to be undignified. Being undignified is what invites abuse and exploitation, because lack of dignity invites belittling, indifference and contempt' (Sollund, 2014).

In need of a deeper acknowledgment of harm

Through Patarroyo's practices, a lot of suffering and damage was perpetrated, and no reduction in the malaria occurrence rates was achieved. Law allows it, and culture denies its harms. Its justification stands on the blind faith in progress, seen as a unitary process arrogated by positivist science and its technologies where the only subject of rights is the human being, and the continuation of the harms in question is fueled by the inequalities of capitalism that refuse to care about the systematic threats that the existence of the impoverished and marginalized. A deeper acknowledgment of harm/wrongdoing is required. Harmful practices continue and a shallow legal acknowledgment has proved to be insufficient.

A longstanding view within critical criminology is that because of its political and ideological constitution, law is not and has never been a good scale by which to measure which events deserve acknowledgment (e.g. Baratta, 2004). A harm perspective seems more fitting for this task (Hillyard et al., 2004). Harm is the stifling prevention or diminishing of potentials 'by systematic pressures and limits of the society as a whole [whereas] the prime goal of any society should be to maximize the enjoyment of dominion' (White, 2013).

Colombian environmental law, based on the concepts of natural resources and sustainable development as supported by the modern ideology of progress, sustains harm. If the aim is to put an end to the diminishing of potentials of non-humans, ecosystems and humans caused by systemic pressures (White, 2013) that stem from human ways of relating to the environment, an alternative yardstick for acknowledging the damage of certain human practices is required. White (2013) proposed eco-justice, which he understands as being the best way 'to achieve justice by ensuring that potential ... is realized as far as practicable [whereas] suffering and degradation [are] diminished'. Others (see, e.g., Sollund, 2008; Svärd, 2012; Beirne, 1999; Cazaux, 1999) have implicitly proposed what has recently been termed the 'species justice perspective', which means caring for the intrinsic value of all individuals, regardless of their species. Whereas these two perspectives can be applied simultaneously, species justice is a prerequisite for eco-justice as one can only speak of justice 'if each individual of that species does not suffer from human inflicted abuse' (Sollund, 2013b).

Conclusion

It is not a scientific position but a life stance. Animals were here before humans, and they have a power over territory, they are powerful, they carry energy that protects them. The relationship to the environment must be consensual. Not between a subject and an object, but among subjects. For all us to live on earth peacefully we need to make agreements with them. We do not care about them because they are cute, but because they have rights and because they own the territory. Because they are powerful they can provide nourishment, but they can also generate sickness. We are interdependent for our survival. If you are going to cut part of a forest owned by animals, then you must restore what you took.[20]

Denying that abducting animals is harmful has led to great damage. Behind these events lies an ideology that allows us to exploit them while denying the resulting harm to all life. If we value animals for the benefit that they offer us, but not for the intrinsic value that they possess – as capitalist culture has taught us (Ferrell et al., 2012) – harms will continue with impunity.

As illustrated in this case study, a yardstick departing from the modern ideology of progress has failed to provide an effective measurement scale to protect the most basic needs/rights of both humans and non-humans. A new one is required: both eco-justice and species justice can fill that space. No contradiction exists between them. Recognizing the intrinsic rights of non-humans does not mean that those rights will always be untouchable, but that we respect their holders by not treating them as commodities, resources or property (Francione, 2008). As with human rights, depending on circumstances, certain rights will have to yield to others, as there is not a single suitable solution to every issue. Furthermore, species justice is a perspective that not only focuses on each individual's rights, including those of the human species, but also contains the critiques of speciesism (Sollund, 2008) and androcentrism (Sollund, 2012b), that have the power to dismount the denials of others' (Young, 2007) intrinsic worth. Such tragic violations against nature and impoverished communities will continue to occur if we do not embrace an alternative yardstick.

Acknowledgement

I am grateful to Ragnhild Sollund, Hanneke Mol and Héctor Chamorro for valuable comments on this chapter.

Notes

1. The acid attack, also known as vitrolage, is a method of assault where the perpetrator throws acid on the victim with the intention of causing physical and moral damage. The physical harm involves third-degree burns. In Colombia, by 2013, around 60% of such attacks have been explained as cases where men were seeking to control women.
2. Of interest for this chapter is the idea that experimentation such as that suffered by non-human animals will not happen to us, human beings. However, history and literature tells us otherwise. See, for example, Skloot (2010, Chapter 16).
3. How to refer to animals other than humans has always been problematic because, as Sollund points out, the term 'animal' 'contrasts the human species to other animals as though they were different in aspects which are important in attributing rights or capacities' (2013b: 2). Terms such as 'non-human animals' or 'animals other than humans' are equally problematic because they create an even sharper contrast with humans, with a connotation of lacking human qualities or virtues. For the sake of simplicity I will refer to animals other than humans with the criticizable term of 'animals'.
4. *Aotus* is a genus of primates (owl monkeys) that are widely distributed across southern Central America and northern South America. They and the human species share a similar immunological system. Because of this characteristic they are extensively 'used in medical research' (Morales-Jiménez et al., 2008).
5. Malaria is a parasitic disease that is transmitted when an infected mosquito bites a human being, injecting them with a parasite larva that goes directly to the human's liver. There it reproduces at a rate of 30,000 times per day.

Once in the body, each infective larva contaminates red blood cells that will later break, releasing 50 new parasites into the body every two days. The malaria parasite causes, among other things, diminished transportation of oxygen and organ failure. Symptoms of the disease include fever, vomiting, headaches and chills. If untreated it can progress to severe illness, such as anaemia or cerebral malaria, and finally death. According to the WHO, in 2012 there were 207 million cases of malaria, of which 627,000 ended in death (WHO, 2013b).

6. Between 26 April 2014 and 4 July 2014, I conducted four interviews for this chapter, all of them with people who were directly involved in the Patarroyo case. Three of the interviewees agreed to be publicly identified, while the fourth asked to remain anonymous for security reasons. To gain an insight into Manuel Elkin Patarroyo's views, I attended public hearings concerning his case. In addition I went to scientific meetings where biologists, primatologists and vets debated the alleged harms/benefits of Patarroyo's research.

7. The protozoan parasite that causes malaria in humans.

8. The Colombian environmental administrative configuration was transformed on 4 July 1991 with the promulgation of a new political constitution. Before it, 'natural resources' were administered by the Colombian National Institute of Natural Resources (INDERENA), which was dependent on the Ministry of Agriculture. Following the launch of the 1991 constitution, the new entity in charge of environmental issues is the Ministry of the Environment. What used to be centralized management of the environment is now a decentralized system, where regional autonomous corporations are in charge of administering each ecosystem's environment. There are 34 Regional Autonomous Corporations in Colombia. Corpoamazonía is in charge of the Amazonian ecosystem in the country.

9. No. 4 in Article II of the CITES convention states: 'The Parties shall not allow trade in specimens of species included in Appendices I, II and III except in accordance with the provisions of the present Convention.' As *Aotus nancymaae* is listed in Appendix II of CITES as exclusively belonging to Brazil and Peru, its introduction into Colombian territory requires a permit, since Article IV of the same convention states: 'The export of any specimen of a species included in Appendix II shall require the prior grant and presentation of an export permit.'

10. Articles 329 (illicit use of renewable resources) and 330A (illicit use of exotic species) of the Colombian Criminal Law.

11. Quote taken from a speech that Manuel Elkin Patarroyo gave at a public hearing that was held in the Colombian State Council, as part of the procedures that were established to solve the lawsuit filed by him against a ruling of the same corporation. In it he explained his view of the issue (including the biomedical considerations), trying to persuade the magistrates to overturn the decision that annulled his hunt and research permits, which allowed him to abduct owl monkeys.

12. He along with Paul Bloor presented the results of the project, 'Study of the Aotus genus at the south of the Colombian Amazonia', funded by the Colombian Ministry of the Environment, on 4 July 2014 at the National University of Colombia.

13. The Amazon basin has been a historically and systematically exploited region. By the 1850s there were the rubber plantations, by the 1960s there were the hunting missions to catch jaguars and tigrillos because of the cat-skin boom (known as the tigrilladas), and currently it is a region of intense trafficking in drugs, arms, non-human animals and human beings (see Vargas Velásquez et al., 2010).

14. See note 11.

15. Authors such as Barreda (2002) have asked for the replacement of the concept of bioprospecting with biopiracy. Bioprospecting can be defined as all of the research, transformation, commercialization and patenting activities related to biodiversity whose goal is to develop products for commercialization, and where systematic study, transformation, commercialization and protection by the patent system of biological-genetic resources are implied (Rodríguez Goyes, 2013b). For them the concept of bioprospecting, by using juridical terms, hides the harmful effects of these practices, while the latter explicitly denounces the extraction (called 'looting' by them) by international enterprises of the natural richness of 'undeveloped' countries.

16. On 29 July 2014, the last news in this regard was released. BBC News reported that GlaxoSmithKline had developed and patented a malaria vaccine with 80% efficiency and offering 18 months of protection. It is awaiting approval for global use (Mundasad, 2014). Let us hope that history does not turn this into another malaria carcass and that it is made accessible to marginalized communities.

17. Authors' translation from the Spanish.

18. The events were reconstructed based on a personal interview with Ángela Maldonado, as well as on the rulings of an administrative tribunal in Colombia and the Colombian High Council.

19. Even when more ecocentric positions may exist within environmental justice, I argue that it is impossible to relate to nature in a non-biased and exploitative logic that departs from a consideration of nature as a resource to satisfy human needs/desires.

20. Description made to me by Clara Henao of the stance that is held by Amazonian aboriginal communities in their relationship with nature.

References

Acción Popular (2013) Demandante: Ángela María Maldonado Rodríguez y otros Demandando: Ministerio de Medio Ambiente, Fundación Instituto de Inmunología de Colombia y otros. *Gil Botero, Enrique.* Consejo de Estado, Sala de lo Contencioso Administrativo, Sección Tercera, Subsección C.

Agnew, R. (1998) 'The causes of animal abuse: A social-psychological analysis', *Theoretical Criminology,* 2: 177–209.

Baratta, A. (2004) *Criminología crítica y crítica al derecho penal.* Buenos Aires: Siglo Veintiuno editores.

Barreda, A. (2002) 'Biopiratería, bioprospección y resistencia: cuatro casos en méxico', *El Cotidiano,* 18: 119–144.

Becker, H. (2010) *Outsiders, hacia una sociología de la desviación.* Buenos Aires: Siglo veintiuno editores.

Beirne, P. (1999) 'For a nonspeciesist criminology: Animal abuse as an object of study, *Criminology*, 37: 117–148.

Brown, P. (1999) Scientist whose dream of beating disease came true. *Guardian*, 24 July.

Bueno Angulo, M.L. and Brieva Rico, C. (2007) Concepto técnico sobre el documento titulado 'Estimación del status actual de las poblaciones naturales de micos del género aotus en San Juan de Atacuari en el trapecio amazónico colombiano'. Bogotá.

Cariboni, D. and Viera, C. (2007) 'Salud-Colombia: La malaria canta victoria en el Chocó, *Inter Press Service*, 29 October.

Cazaux, G. (1999) 'Beauty and the Beast: Animal abuse from a nonspeciesist criminological perspective', *Crime Law and Social Change*, 31(2): 105–126.

Cohen, S. (2001) *States of Denial: Knowing about Atrocities and Suffering*. Cambridge: Polity Press.

Cohen, S. (2011) *Folk Devils and Moral Panics: The Creation of the Mods and Rockers*. Oxon: Routledge.

Cohen, S. (2013) 'Panic or denial, on whether to take crime seriously'. In Gilmore, J., Moore, J.M. and Scott, D. (eds) *Critique and Dissent: An Anthology to Mark 40 Years of the European Group for the Study of Deviance and Social Control*. Ottawa: Red Quill.

Del-Medio-Ambiente, M. (2000) 'Decreto 309 de 2000'. In Colombia, P.D.L.R.D. (ed.) Colombia. Accessed at http://www.alcaldiabogota.gov.co/sisjur/normas/Norma1.jsp?i=45528.

Desowitz, S.R. (1991) *The Malaria Capers*. London: W.W. Norton & Company.

Desowitz, S.R. (1997) *Who Gave Pinta to the Santa María. Tracking the Devastating Spread of Lethal Tropical Diseases into America*. New York: W.W. Norton & Company.

Desowitz, S.R. (2002) *Federal Bodysnatchers and the New Guinea Virus*. New York: Norton.

Ferrell, J., Hayward, K. and Young, J. (2012) *Cultural Criminology, an Invitation*. London: Sage Publications.

Francione, G.L. (2008) *Animals as Persons: Essays on the Abolition of Animal Exploitation*. New York: Columbia University Press.

Goffman, E. (1997) *La presentación de la persona en la vida cotidiana*. Buenos Aires: Amorrotu Editores.

Graves, P.M. and Hellen, G. (2006) 'Vaccines for preventing malaria (SPf66)', *Cochrane Database of Systematic Reviews*, 2: CD005966–CD005966.

Hall, S. (1997) *Representation: Cultural Representations and Signifying Practices*. London: Sage Publications.

Hernández Álvarez, M. and Obregón Torres, D. (2002) *La OPS y el Estado colombiano: Cien años de historia 1902–2002*. Bogotá, Organización Panamericana de la Salud.

Hernández Álvarez, M. and Torres-Tovar, M. (2010) 'Nueva reforma en el sector salud en Colombia: portarse bien para la salud financiera del sistema', *Medicina Social*, 5: 241–245.

Hillyard, P., Pantazis, C., Thombs, S., Gordon, D., Salmi, J., Pemberton, S., Ward, T., Naughton, M., Sim, J., Webber, F., Dorling, D., Bibbings, L., Parker, R. and Gordon, D. (2004) *Beyond Criminology, Taking Harm Seriously*. London: Pluto Press & Fernwood Publishing.

Hoyos Vásquez, G. (2006) Ciencia y ética desde una teoría discursiva. *VI jornadas latinoamericanas de estudios sociales de la ciencia y la tecnología*. Bogotá, Colombia.

Hulsman, L. (2003) Conflictos relativos a la terminología: 'Situación problemática' vs. 'Crimen'. *Cátedra de Investigación Científica*. Bogotá: Universidad Externado de Colombia.

Hulsman, L. and Bernat De Celis, J. (1984) *Sistema penal y seguridad ciudadana: Hacia una alternativa*. Barcelona: Editorial Ariel.

IUCN/SSC (2013) *Guidelines for Reintroductions and Other Conservation Translocations*. Gland, Switzerland, IUCN Species Survival Commission.

Lynch, M.J. (2006) 'The greening of criminology: A perspective for the 1990s'. In Beirne, P. and South, N. (eds) *Green Criminology*. Hampshire, UK: Aldershot.

Maldonado, Á.M. (2014) Informe para el Comité de Verificación sobre los fallos en primera y segunda instancia proferidos por el Tribunal Administrativo de Cundinamarca y el Consejo de Estado, Sección Tercera en la Acción Popular 2011–227.

Maldonado, Á.M. and Peck, M.R. (2014) 'Research and in situ conservation of owl monkeys enhances environmental law enforcement at the Colombian-Peruvian border', *American Journal of Primatology*, 76(7): 658–669.

Mol, H. (2013) ' "A Gift from the Tropics to the World": Power, harm and palm oil'. In Westerhuis, D., Walters, R. and Wyatt, T. (eds) *Emerging Issues in Green Criminology: Exploring Power, Justice and Harm*. New York: Palgrave Macmillan.

Morales-Jiménez, A.L., Link, A., Cornejo, F. and Stevenson, P. (2008) *Aotus vociferans* , The IUCN Red List of Threatened Species, http://www.iucnredlist.org/details/41544/0. Accessed on 3 August 2014.

Mundasad, S. (2014) ' "Milestone" for child malaria vaccine'. *BBC*, 29 July.

Nietzsche, F. (2005) *La genealogía de la moral*. Madrid: Alianza Editorial.

Paz-Y-Miño, C. (1997). Biodiversidad y bioprotección en genética humana. In Varea, A. (ed.) *Biodiversidad, bioprospección y bioseguridad*. Quito: ILDIS; Instituto de Estudios Ecologistas del Tercer Mundo; FTPP-FAO, ABYA-YALA.

Rodríguez Goyes, D. (2013a) Dimensions of conflict and punishmen. An Analysis of the acid attacks towards women in Colombia. *41 annual conference of the European Group for the Study of Deviance and Social Control*. Oslo.

Rodríguez Goyes, D. (2013b) Introducción general, las determinantes como fenómenos sociales a considerar en la construcción de políticas públicas para la bioprospección en Colombia. In Melgarejo, L.M. and Toro pérez, C. (eds) *Determinantes Científicas, Económicas y Socio-Ambientales de la Bioprospección en Colombia*. Bogotá: Universidad Nacional de Colombia, pp. 19–42.

Schlosberg, D. (2007) *Defining Environmental Justice. Theories, Movements, and Nature*. New York: Oxford University Press.

Skloot, R. (2010) *The Immortal Life of Henrietta Lacks*. New York: Crown Publishing.

Sollund, R. (2008) 'Causes for Speciesism: Difference, Distance and Denial'. In Sollund, R. (ed.) *Global Harms: Ecological Crime and Speciesism*. New York: Nova Science Publishers, Inc.

Sollund, R. (2011) 'Expressions of speciesism: The effects of keeping companion animals on animal abuse, animal trafficking and species decline', *Crime Law and Social Change*, 55(5): 437–451.

Sollund, R. (2012a) Speciesism as Doxic Practice versus Valuing Difference and Plurality. In Ellefsen, R., Sollund, R. and Larsen, G. (eds) *Eco-global Crimes, Contemporary Problems and Future Challenges.* Surrey: Ashgate Publishing Limited.

Sollund, R. (2012b) 'Victimisation of women, children and non-human species through trafficking and trade: Crimes understood under an ecofeminist perspective'. In South, N. and Brisman, A. (eds) *Routledge International Handbook of Green Criminology.* Londres: Routledge International Handbooks.

Sollund, R. (2013a) 'Animal abuse, 'wildlife' trafficking and speciesism'. In Bruinsma, G. and Weisburd, D. (eds) *Encyclopedia of Criminology and Criminal Justice.* Tasmania: Springer.

Sollund, R. (2013b) 'Animal abuse, animal rights and species justice'. *American Society of Criminology 69th Annual Meeting.* Atlanta: American Society of Criminology.

Sollund, R. (2013c) 'Animal trafficking and trade: Abuse and species injustice'. In Walters, R., Westerhuis, D. and Wyatt, T. (eds) *Emerging Issues in Green Criminology.* London: Palgrave Macmillan.

Sollund, R. (2014) 'A comment to Gary Francione: Animal rights versus animals as property and nature'. In Cederholm, E., Björck, A., Jennbert, K. and Lönngren, A.-S. (eds) *Exploring the Animal Turn.* Lünd: Pufendorf.

Stake, R. (2005).' Qualitative case studies'. In Denzin, N. and Lincln, Y. (eds) *The Sage Handook of Qualitative Research,* 3rd ed. London: Sage Publications.

Stretesky, P.B., Long, M.A. and Lynch, M.J. (2014) *The Treadmill of Crime: Political Economy and Green Criminology.* New York: Routledge.

Svärd, P.-A. (2012) 'The ideological fantasy of animal welfare: A Lacanian perspective on the reproduction of speciesism'. In Ellefsen, R., Sollund, R. and Larsen, G. (eds) *Eco-global Crimes, Contemporary Problems and Future Challenges.* Surrey: Ashgate Publishing Limited.

Sykes, G.M.S. and Matza, D. (1957). 'Techniques of neutralization: A theory of delinquency. *American Sociological Review,* 22: 664–670.

Vargas Velásquez, A., Álvarez Gómez, C., García Pinzón, V., González Cely, L.J., Ortega Gómez, A.F., Pabón Ayala, N., Sánchez García, D.P. and Vera Arias, A.M. (2010) *Inseguridad en la Región Amazónica.* Bogotá: Universidad Nacional de Colombia.

Vattimo, G. (1990) *La sociedad transparente.* Barcelona: Ediciones Paidós.

Wallner, F. (1994) *Constructive Realism, Aspects of a New Epistemological Movement.* Viena: Wilhelm Braumüller.

White, R. (2012) 'The foundations of eco-global Criminology'. In Ellefsen, R., Sollund, R. and Larsen, G. (eds) *Eco-global Crimes, Contemporary Problems and Future Challenges.* Surrey: Ashgate Publishing Limited.

White, R. (2013) *Environmental Harm: An Eco-justice Perspective.* Bristol: Policy Press.

WHO (2013a) *Country Profiles.* France: WHO Publications.

WHO (2013b) *World Malaria Report.* France: WHO Publications.

Young, J. (2007) *The Vertigo of Late Modernity.* London: Sage Publications.

10

A Systems Thinking Perspective on the Motivations and Mechanisms That Drive Wildlife Poaching

Joanna F. Hill

> What constitutes an explanation of an observed social phenomenon? Perhaps one day people will interpret the question, 'Can you explain it?' as asking 'Can you grow it?'
>
> (Epstein and Axtell, 1996: 20)

Introduction

Poaching[1] is a type of wildlife crime that is becoming increasingly problematic in many countries (Harrison, 2011). It has been linked to the spread of disease between non-human animals and people (Swift et al., 2007), threatens endangered species with extinction (Bouché et al., 2010), and has been connected to other forms of crime, such as weapons and drugs smuggling (Ayling, 2013; South and Wyatt, 2011). Consequently there has been a growing desire to understand the mechanisms that drive poaching in order to develop effective solutions (see von Essen et al., 2014).

Although a critical analysis of poaching would befit this volume (e.g. see Büscher et al., 2012; Moore, 2011), instead I analyse poaching as an empirically trained researcher with a growing interest in the cultural and political contexts that define (and 'solve') environmental crime. While critical criminology is important, the empirical side of understanding and reducing environmental crime should not be neglected. Therefore, drawing from ongoing doctoral research on wildlife poaching, I explain why a 'systems thinking' and computer simulation modelling approach can be used by both critical and empirical researchers to develop their understanding of environmental crime. Systems thinking adopts a holistic approach to analysing complex systems, particularly by focusing upon how the interactions between individuals might generate complex phenomena (Jackson, 2003). Computer simulation models are often utilized to explore complex systems within an artificial environment and to test different policies that would be difficult to achieve in reality for financial, logistical or ethical reasons (Gilbert and Troitzsch, 2005).

This chapter is divided into seven sections. In the first, some commonly used terms and definitions are discussed. The second describes a working poaching taxonomy to better understand the different motivations for someone to poach. Instrumental and normative mechanisms that drive poaching are reviewed in the next two sections, followed by a brief overview (in sections five and six) of systems thinking and simulation perspectives, and how these might relate to poaching. Finally, the seventh section summarizes the key points and considers some future directions for this research.

Defining and delimiting wildlife poaching

Before reviewing the different motivations and mechanisms that drive wildlife poaching, some commonly used terms should be briefly described. 'Wildlife'[2] includes all forms of non-domesticated animals and plants living in the wild (Lemieux, 2014: 2). 'Wildlife hunting' refers to the legal pursuit of animals or plants for some purpose while 'wildlife poaching' refers to the illegal taking of wildlife (Musgrave et al., 1993).[3] 'Poaching' can include the killing or removal of an alive or dead individual for some purpose (e.g. for the pet trade or personal consumption), or their parts/derivatives (e.g. tusks, horns, fins, antlers, eggs, petals, roots and sap) (Moreto and Clarke, 2011). Poaching often (but not always) occurs in 'protected areas', which the IUCN defines as:

> A clearly defined geographical space, recognised, dedicated and managed, through legal or other effective means, to achieve the long-term conservation of nature with associated ecosystem services and cultural values.
>
> (Dudley, 2008: 8)

Poaching is particularly problematic in protected areas for at least three reasons: (1) the large size of protected areas, makes the monitoring of illegal activity logistically difficult (Dudley, 2008);[4] (2) the fact that they are managed in such a way as to attract wildlife (Okello et al., 2005), which increases the number of potential targets; and (3) their close proximity to urban areas means that wildlife can be quickly and easily accessed and sold (Mcdonald et al., 2009). However, because wildlife also reside outside protected areas where little enforcement exists, there is a need to develop effective poaching-reduction strategies that will have a wider impact (Lahm, 2001).

Recently, Moreto and Clarke (2011) divided poaching into separate types of wildlife crime, as illustrated in Figure 10.1. 'Illegal hunting' (the illegal pursuit of wildlife for some purpose) refers to all hunting activities leading up to (but not including) the point of unlawfully taking an organism from its environment. For instance, 'active hunting methods' involve the pursuit of animals with dogs, spears and guns (Hitchcock, 2000), while 'passive hunting methods' include the setting of traps on animal trails or poisoning water or food supplies (Noss, 1998). Poaching does not occur until a would-be offender comes into direct contact with their target, at which point it refers to the process of illegally taking a plant or animal from the wild at a particular place and time (Moreto and Clarke, 2011). The distinction between poaching and illegal hunting is important because one does not necessarily require the other. For example, although poaching most often occurs as the end result of illegal hunting, opportunistic poaching could also occur in situations where a person happens to discover an unguarded plant or animal and decides to take it. Moreover, illegal hunting may not necessary result in the removal of wildlife. For example, an animal may elude capture or a person may seize an animal but then return it to the wild (such as occurs in catch-and-release sport fishing), or a person may simply hunt for the thrill of the chase. For example, drag hunting is a sport, which involves riding horses alongside packs of hounds to follow an artificially laid scent over a predetermined area; no animals are killed, however (Ward, 1999).

All activities that take place after a plant or animal has been poached represent different types of wildlife crime, which will vary depending upon the end goal of the offender (Moreto and Clarke, 2011). For example, a person might kill a crop-raiding animal and then bury the carcass to try to remove the evidence of their clandestine activity (Hamilton and Erhart, 2012). As another example, if the goal of a person is to obtain money, then once a particular target has been poached, it will often be required to be processed and transported to some location. Depending upon the species and size of the catch, the desired part of the animal might be cut out or off (e.g. horns or tusks), or the edible parts of the animal might be smoked or butchered into smaller pieces and the rest of

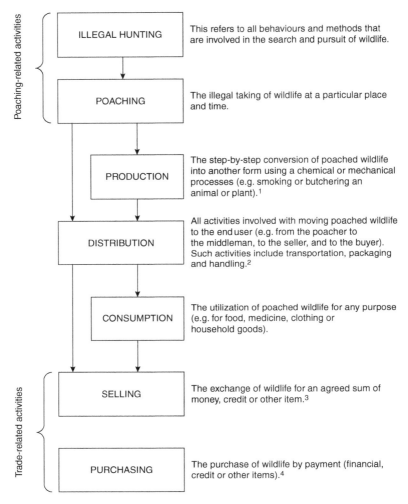

Figure 10.1 A conceptual model adapted from Moreto and Clarke (2011) that illustrates the different stages of illegal hunting, poaching and the wider illegal wildlife trade. The arrows show some possible combinations of the poaching process. For example, a person may process an animal on site before transporting it back to the village for consumption. Or they might transport their catch directly to some location (such as a market for trade)

Notes: 1 Kumar and Suresh (2003: 3). 2 Arnold (1998). 3 *Collins Online Dictionary*. 4 Ibid.

the animal discarded (Coad, 2007). Once the target has been processed it can be transported back to a person's home location (Loibooki et al., 2002) and/or traded at local markets or to other middlemen (Ayling, 2013).

Mapping out the poaching process as illustrated in Figure 10.1 – or the 'poaching script' – is useful for developing poaching-reduction solutions that can be targeted at each stage of the chain to disrupt the activity (for 'crime scripts', see Borrion, 2013; Cornish, 1994).

One key question, however, is should a person be prosecuted for poaching if they had not intended to poach? Muth and Bowe (1998) argue that poaching must be *an intentional* act; if someone accidentally kills an animal or was unaware that they were breaking the law, then they should not be classified as a poacher, although they may still be prosecuted under other wildlife-protection laws. However, conscious intent and genuine ignorance of the law are difficult to determine without being able to directly access someone's thoughts (Eliason, 1999), and this 'ignorance' excuse has been used by offenders to protest their innocence and escape conviction (see Eliason, 2004).

To this end, Moreto and Clarke (2011) draw upon principles from the bounded rationality perspective (see Simon, 1990). According to rational choice theory, when people are faced with a decision between multiple options, they will tend to choose the most optimal solution that balances the benefits against the costs (Becker, 1968; Clarke and Cornish, 1985). However, humans are in fact more likely to make satisficing or satisfactory decisions when faced with multiple choices due to limitations that are associated with their knowledge of the environment, cognitive/perceptual ability, experience and skill or as a result of other social concerns (Simon, 1990). From this perspective, then, Moreto and Clarke (2011) argue that people who poach should be considered as 'bounded rational offenders' who make imperfect decisions on when, where and how to poach wildlife. The implication for this interpretation is that regardless of the reasons why a person poaches (i.e. whether by accident, ignorance of the law or as an intentional act), they should be prosecuted for poaching.

Nevertheless, the methods and motivations for poaching will vary both in degree and in kind depending upon the social, political, cultural and environmental context in which the problem occurs. For example, a common reason for poaching with snares in developing countries is to provide income and food for families due to a lack of alternative livelihood (Kahler and Gore, 2012). On the other hand, the deliberate poaching of top predators with poison or guns to protect game animals is common in many developed countries (Guitart et al., 2010). Consequently it is possible that solutions to poaching may be more or less effective in different contexts (Muth and Bowe, 1998). Although an exhaustive review is beyond the scope of this chapter, the next section briefly describes the different reasons why people might poach by drawing upon some previous research on poaching taxonomies.

Motivations for poaching

The mainstream media often depicts poaching as a battle between selfish and violent poachers who are financially motivated, and environmentalists who are trying to save wildlife from extinction (Duffy, 2014). Some scholars have suggested that this image has facilitated the rapid militarization and advancement of 'shoot-on-sight' policies in protected areas (Duffy, 2014; Neumann, 2004; Wall and McClanahan, 2015). While enforcement strategies can be highly effective in many cases (Hilborn et al., 2006; Messer, 2010), questions are being raised regarding the moral justification for this approach, particularly as many people who poach do so due to a lack of livelihood alternatives (Duffy, 2014). In fact, 20 years of research in the social and conservation sciences has shown that 'poachers' are not a financially motivated homogenous group; rather, people will poach for a large number of complex reasons that vary based on time, place and circumstance (see Moreto & Clarke, 2011; Muth and Bowe, 1998; von Essen et al., 2014).[5]

One of the most systematic and comprehensive classification studies of poaching motivations that has been published to date has been undertaken by Muth and Bowe (1998), who used content analysis of published sources to identify ten different types of poaching: 'household consumption'; 'commercial gain'; 'recreational satisfaction'; 'trophy poaching'; 'thrill killing'; 'gamesmanship'; 'protection of self and property'; 'poaching as rebellion'; 'poaching as traditional right'; and 'disagreement with specific wildlife regulations'. Their examples of poaching focused on studies that were conducted in North America. However, these may not reflect the reasons for or circumstances behind poaching in other parts of the world.

With this in mind, Moreto and Clarke (2011) developed a modified and more inclusive poaching taxonomy from a so-called 'instrumental' perspective, which focuses on how environmental and social factors create opportunities for poaching (see below). They also used content analysis of published literature and drew from a range of countries to

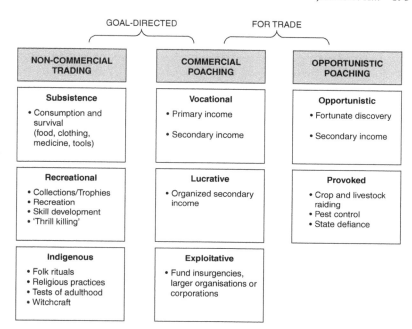

Figure 10.2 A working poaching taxonomy based upon an 'instrumental' criminology perspective. Adapted from Moreto and Clarke (2011)

identify eight reasons for poaching. A taxonomy based upon their findings is illustrated in Figure 10.2, and this has been further classified by whether a person is motivated to poach for commercial reasons, and if poaching is goal-directed (deliberately and consciously conducted for a specific purpose) or opportunistic (undertaken based on the immediate situation or circumstances) (Moreto and Clarke, 2011).

Financially driven poaching

Commercial or financial gain is one of the most common reasons for poaching, particularly in developing countries such as Africa (Lindsey et al., 2012). For example, the 'vocational poaching' of bushmeat (any wild animal hunted for meat) can be a primary source of income for some people (see Loibooki et al., 2002). In many more cases, however, vocational poaching serves as a secondary form of income to supplement agricultural activities, to provide a safety net in the absence of other paid employment or to fund a specific purpose, such as to send a child to school (Coad et al., 2010). In contrast to vocational poaching, 'exploitative poaching' refers to individuals who are hired to target and successfully procure parts of specific animals (e.g. rhinos or elephants) (Moreto and Clarke, 2011; Warchol and Johnson, 2011). Similarly, 'lucrative poaching' refers to larger groups or organizations that target specific species or groups of animals for financial gain. For example, poaching can be used to provide food and resources to fund insurgencies (Dudley et al., 2002), or larger criminal gangs may undertake poaching as part of other illegal activities, such as drug and weapons smuggling (South and Wyatt, 2011). This can also include the large-scale industrial poaching of timber (Tacconi, 2007), and illegal, unreported and unregulated fishing by corporations (Gallic and Cox, 2006).

Opportunistic poaching

Drawing from an instrumental perspective, 'opportunistic poaching' can occur in situations where a person has no prior intention to poach but on encountering an unguarded animal during the course of their daily routine, decides to take it if the perceived benefits outweigh the perceived risks, such as apprehension by rangers (Moreto and Clarke, 2011; Pires, 2012). Alternatively, an individual may poach an animal after discovering it has attacked their domestic animals or raided their crops (Distefano, 2005; Sollund, 2015), or as a means of defying authority in retaliation against conservation strategies that are seen as taking priority over the concerns of local people (Holmes, 2007; Nurse, 2013; Woods et al., 2012). In all of these situations a person may have had no previous intention to poach but a particular immediate situation or some other form of conflict has generated a suitable reason or opportunity for poaching. A poached plant or animal can then be used for personal consumption and/or traded for money or other goods.

Non-commercial poaching

Non-commercial poaching can be goal-orientated, but not necessarily motivated by profit-making purposes. For example, the goal of 'subsistence poaching' is to provide meat or plants for medicine, food or other household products – also known as 'hunting for the pot' (Duffy, 2014; Hitchcock, 2000). 'Recreational poaching' may be undertaken by those who enjoy the thrill of pursuing a target, or the excitement of trying to outsmart park rangers and commit a deviant act. Recreational poaching may also occur out of a desire to develop and test the skills of the person poaching or to acquire trophies (Filteau, 2012; Forsyth and Marckese, 1993). In addition, it may include people who enjoy collecting particular wild plants or animals, such as illegal egg collectors, which has been likened to a form of obsessive-compulsive disorder whereby people compete to assemble the biggest and best collections (see Nurse, 2013: 137). Finally, 'indigenous poaching' refers to any poaching for sociocultural reasons (McCorquodale, 1997). Animals and plants have been, and continue to be, hunted for festive holidays, dowries, medicine, or for use in initiation ceremonies or other tests of adulthood (Kideghesho, 2009). In some cultures the hunting of particular species is taboo, which Saj et al. (2006: 289) define as 'an unwritten rule or prohibition that regulates human behavior, by causing avoidance or restraint'. The extent to which people adhere to a taboo will often depend upon their fear of the supernatural or the social consequences of violating the rules (Kideghesho, 2009).

It should be noted that these categories of poaching are not necessarily mutually exclusive; they can be interconnected and context dependent. For example, a person might poach if an opportunity arises (e.g. on discovering an unguarded plant or animal), but their motivation might switch to vocational poaching in order to provide additional meat for a festival (Mendelson et al., 2003), or during the dry seasons when agricultural activities can no longer provide enough income for families (Lindsey et al., 2012).

In summary, Moreto and Clarke's (2011) taxonomy provides a refreshing and systematic means of better understanding poaching. It is clear that people poach for a number of reasons, which may have implications for developing context-specific solutions (Muth and Bowe, 1998). For example, digging trenches and planting chilli plants may help to deter the poaching of crop-raiding animals (Davies et al., 2011). However, it would be ineffective and unfeasible to deploy the same strategy across an entire protected area to deter people who are motivated by hunger and poverty. Furthermore, it is possible that the blanket use of enforcement and shoot-on-sight policies could unintentionally radicalize people into poaching and thus escalate the problem further (von Essen et al., 2014). To better understand what creates these different motivations, the following two sections outline different mechanisms that drive poaching from an 'instrumental' followed by a 'normative' perspective.

Mechanisms that drive poaching

Two perspectives which have been put forward hitherto to explain poaching have focused on either 'normative' explanations, which explore how norms and values influence a person's willingness to offend (Eliason, 1999), or 'instrumental' explanations, which focus upon how an individual responds to the immediate situation and their perceived incentives and costs of committing a crime (Becker, 1968; Cohen and Felson, 1979). These are discussed below.

Instrumental mechanisms that drive poaching

Instrumental or 'opportunity' perspectives[6] of crime focus on crime events and how the immediate social and environmental situation can create crime opportunities (Brantingham and Brantingham, 1981; Clarke and Cornish, 1985; Cohen and Felson, 1979). For instance, according to the 'routine activity perspective' (Cohen and Felson, 1979), the probability of a crime event occurring (e.g. a house being burgled or an animal being poached) is a function of the convergence of at least three sets of actors: when a motivated offender (a person that burgles homes or poaches animals), encounters a suitable target (a house or an animal) in the absence of a capable guardian (a CCTV camera, a park ranger or even a dangerous animal that could harm a person)

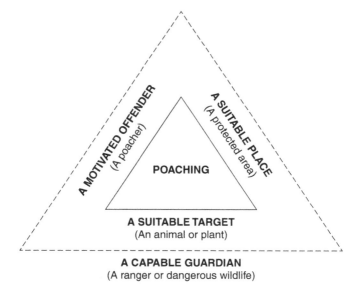

Figure 10.3 An example of a 'poaching crime triangle', illustrating how the convergence and interactions of offenders, targets and guardians can lead to a crime (poaching) event occurring

at a particular place and time. Figure 10.3 shows a crime triangle of poaching, which is often used by crime scientists to show the important interactions in the formation of a crime event (Clarke and Eck, 2005).

According to the routine activity perspective, most people (offenders and non-offenders) have regular behavioural routine activities that they engage in to meet their basic needs (Cohen and Felson, 1979). For example, people travel from their homes to work at certain times in order to generate income, they visit shops to purchase food and they go to recreational areas for social interaction. In the case of poaching, many non-human animals also have routine activity patterns. For example, hippopotami (*Hippopotamus amphibius*), which are often targeted by poachers (Olupot et al., 2009), will regularly emerge from the water to graze at night when temperatures are cooler, and then return to the water during the day when temperatures are too hot (Olivier and Laurie, 1974). On the other hand, people who poach may often engage in 'criminal behavioural routines' (as opposed to legitimate routines), particularly in the case of snare poaching, which involves someone regularly searching for optimal places to lay snares and checking snares for trapped animals (Coad et al., 2010).

Brantingham and Brantingham's (1981) 'crime pattern theory' expands these ideas by emphasizing how the particular activity and awareness spaces of offender and victims, and the way in which they intersect, can generate crime opportunities. Figures 10.4A and 10.4B illustrate these concepts by comparing an urban and non-urban setting. Over time, people construct their own 'awareness spaces', which represent particular places of high human activity and paths which are regularly used as people (or animals) go about their daily routine (Brantingham and Brantingham, 1981). Therefore the legitimate routine activity patterns of people should help to shape an offender's awareness of particular places and crime opportunities that they are able to exploit (Brantingham and Brantingham, 1981). Moreover, 'rational choice theory' (Becker, 1968; Clarke and Cornish, 1985) emphasizes that an offender will commit a crime based on whether the perceived benefits of the crime (prestige, experience, thrills, financial gain) outweigh the effort and risks (injury, detection, effort) and their likelihood of succeeding.

Thus the instrumental approach suggests that crime is not random but instead tends to converge in space-time 'crime hotspots' that reflect how the routine activities of offenders and victims in the absence of adequate guardianship generate crime opportunities (Johnson, 2010). There is also evidence to suggest that wildlife poaching clusters in a

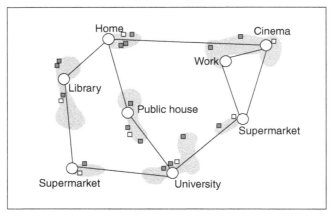

(A) Urban activity spaces

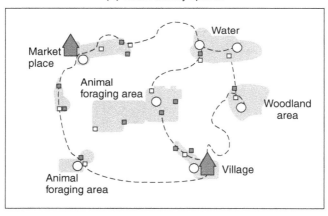

(B) Non-urban poaching activity spaces

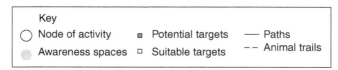

Figure 10.4   Illustration of target choice behaviour by offenders in an urban setting (A) and poaching in a non-urban setting (B). Adapted from Brantingham and Brantingham's (1981) crime pattern theory

similar way, such as around animal trails, water points and park borders (Wato et al., 2006; Watson et al., 2013). Consequently, manipulating the environment to reduce or disrupt the convergence between offenders and their targets around activity spaces and nodes should, in theory, prevent crime (and poaching) from occurring (Johnson, 2010). These principles and associated spatial crime-mapping techniques have been successfully utilized to study and reduce street crime in urban environments (see Chainey et al., 2008), and there is great potential for applying these techniques to understanding environmental crime (Lemieux et al., 2014).

Nevertheless, opportunity theories have been criticized for sometimes failing to predict criminal behaviour (see Kroneberg et al., 2010). This may be due to their failure to incorporate societal norms and values while also overestimating people's ability to make optimal rational choices (von Essen et al., 2014: 6). Therefore the next section briefly describes five normative mechanisms that may also drive poaching.

Normative and integrated mechanisms that drive poaching

Normative perspectives focus upon the role of morals and socialization that influence whether an individual commits an offence (Eliason, 1999). As Muth and Bowe (1998: 10) state,

> Poaching is often embedded in subcultural webs of meaning that involve tradition, ethnic heritage, individual and social identities, and other socio-cultural factors.

According to the normative perspective, an individual complies with wildlife regulations to the degree that the law is perceived as appropriate, legitimate and consistent with their internalized norms (von Essen et al., 2014). At least three normative mechanisms may be relevant to poaching. First, differential association theory suggests that when people interact with their peers and wider environment, they become socially conditioned with the norms and values to justify their involvement with poaching, while also learning the practical tools to hunt (Eliason, 1999; Sutherland, 1947). Similarly, neutralization theory focuses upon the techniques that offenders use to rationalize, justify and reduce the cognitive dissonance that they feel towards their deviant behaviour (Sykes and Matza, 1957), which are learned from their interactions with deviant others. Some example neutralizations as discussed by Eliason and Dodder (1999), Klockars (1974) and Sykes and Matza (1957) that people have used to justify their poaching activities include the following: (1) an offender claims ignorance of the law or that the poaching incident was an accident (denial of responsibility); (2) they claim that their good qualities makeup for poaching (metaphor of the ledger); (3) they state that they poached for reasons of poverty (defence of necessity); or (4) they deflect the blame by pointing out the corrupt and illegal activities of enforcement (condemnation of the condemners). Any of these tools can be used by people to justify their actions so that they are psychologically able to commit the crime and to justify their actions after the event has occurred (Eliason, 1999; Enticott, 2011). Finally, defiance theory suggests that people may use neutralization techniques and become radicalized into poaching as a consequence of perceived overly harsh treatment by those who wish to prevent poaching (von Essen et al., 2014).

In fact, it is likely that normative and instrumental theories have a role to play in poaching behaviour (Kroneberg et al., 2010; von Essen et al., 2014). For example, Wikström (2006: 70) states:

> A proper theory of action is a theory that specifies the causal processes that link the individual's characteristics and experiences (predispositions) and the features of his environment (inducements and constraints) to his acts.

There are two relevant theories that explain how norms and instrumental mechanisms might interact to cause crime. First, the situational action theory of crime causation' suggests that people who have strong internalized pro-social norms do not consider committing a crime as an option, even when placed in a situation in which it would be easy to commit an offence without detection (Wikström, 2006). Similarly, the model of frame selection states that people refrain from law-breaking if their pro-social norms and morals are strongly internalized, and if the immediate situation does not legitimize norm-breaking (Etzioni, 1988; Sykes, 1978). It is only when these conditions are broken that a person will utilize rational choice models to determine whether or not to commit an offence (Kroneberg et al., 2010; for a review, see Sattler et al., 2013).

Although these more integrated theories have yet to be explored with wildlife crime, norms play a crucial part in determining whether or not a person is motivated to poach (Eliason, 1999). Therefore there is a need to develop relevant theoretical and methodological frameworks that can allow the testing of instrumental and normative theories and their interactions under different contexts and poaching motivations. This could be achieved using integrative offender motivational-situational frameworks, such as the conjunction of criminal opportunity model (for a review, see Ekblom and Tilley, 2000). The next section, however, makes a case for the use of systems thinking and simulation modelling techniques that can be used for such theory development and testing.

Systems thinking and complexity theory perspectives

Systems thinking provides an anti-reductionist approach and set of methods for analysing complex systems (Mattessich, 1982). A key idea is that phenomena cannot be explained by looking at their structural components alone. Rather, knowledge and understanding are developed by looking at the whole system and the interactions between its component parts (Cabrera et al., 2008). Systems thinkers often compare natural and simulated systems using conceptual diagrams, as well as system dynamics and computer simulation models to identify their properties and underlying mechanisms (Meadows, 2008).

There are three approaches behind systems thinking which interpret complex systems very differently – namely, 'hard systems thinking', 'soft systems thinking' and 'critical systems heuristics' (Mingers and White 2010). First, the hard systems thinking approach emerged during advancements in biology, general systems theory (von Bertalanffy, 1950), cybernetics (Wiener, 1948) and complexity science (Simon, 1962). Drawing from a post-positivist worldview, proponents of hard systems thinking assume that systems, such as biological cells, the immune system, ecosystems and communication networks, are real, tangible, 'out there' phenomena. These open systems maintain relative stability (homeostasis) by continually reacting and exchanging information with an ever-changing environment and as a result of other feedback processes (von Bertalanffy, 1950). William (2012: 356) illustrates this with two simple examples of a thermostat and radiator heating control system (negative feedback), and the rapidly increasing sound that is produced when a microphone is placed near a loudspeaker (positive feedback). Positive feedback systems are normally associated with negative consequences as they exponentially amplify an effect that can lead to system instability. For example, changes in forest structure and climatic warming as a result of illegal logging and deforestation can increase a forest's susceptibility to fire, which then becomes more likely and severe with each fire occurrence (Cochrane, 2003).

Researchers working in the field of complexity science have developed these key ideas further by emphasizing the importance of system adaptation, emergence and complexity (Simon, 1962; Weaver, 1948). The term 'complexity' is broadly defined as a system that comprises a large number of interacting components and nonlinear feedback processes (Simon, 1962). Complex systems comprise adaptive and autonomous individuals or 'agents' whose nonlinear interactions at the microlevel can lead to the emergence of complex macrolevel behaviour (Holland,

1992). For example, the complex patterns that are observed in bird flocks (or fish shoals or insect swarms) can be shown not to occur because of an invisible hand that guides the group but because the individuals maintain a preferred distance and angle from their nearest neighbour (Kernick, 2004). Such emergent phenomena are generated by bottom-up mechanisms (e.g. natural selection) rather than through a top-down controller (Bersini, 2012).

During the 1970s there was an epistemological shift to a soft systems thinking approach based on social constructivism (Checkland, 1981). Instead of models being representations of reality, they are utilized as tools for understanding. Proponents of soft systems thinking believe that too much emphasis is placed on biological concepts of structure and function to explain social phenomena, rather than other cultural or political processes (Mingers and White, 2010). Furthermore, social models are often developed for policy development or as a decision-making tool (Thorngate and Tavakoli, 2009). Stakeholders, however, can have very different views about defining a problem, and so any recommendations that are generated by a model can be regarded as not sufficiently or fully addressing a problem and thus unacceptable (Midgley, 2006). For example, in the UK, the implementation of airport X-ray body scanners, which are designed to detect security threats, were met with much unexpected public criticism due to human rights and privacy concerns (Mitchener-Nissen et al., 2012). Failing to account for the social consequences of models could lead to wasted time and money when implementing policy changes in the real world.

A third and final approach is critical systems heuristics, which combines both soft and hard systems thinking into a pragmatic and mixed-method framework based on the research problem (Ulrich, 1983). Critical systems heuristics is useful in situations where stakeholders could be marginalized based on particular decisions that are made as a result of the model (see Midgley, 2006). This is relevant to the field of poaching because the failure of some conservation policies has been blamed, in part, for neglecting the needs of local people (see Duffy, 2010). With a critical systems heuristics approach, participation and collaboration become the important components of the system (or problem) under study because changes to the system must involve all stakeholders with adequate representation of their views (Midgley, 2006). Many of these ideas thus compliment the issues raised by Heckenberg and White (2013), who argue that for research to be ethical and effective in the long term, researchers must critically consider whose interests and voices are to be heard (e.g. animals, communities, park managers and governments), as well as relations of power between various stakeholders. With some modifications, a critical systems heuristics approach could serve as a useful tool for developing more effective and ethical policies for preventing poaching and other kinds of environmental crime in the long term (Barnaud et al., 2010; Larsen, 2011).

Because many complex systems and problems are difficult to analyse using thought experiments or statistical methods alone, computer simulation modelling techniques can be employed to identify their properties and underlying mechanisms, as well as to test different environmental policies on different stakeholders that would not be possible using statistical or ethnographic methods (Meadows, 2008). The basic principles of simulation modelling are briefly described in the next section, with some examples of testing different poaching-reduction strategies.

Simulation modelling

Computer simulation models are artificial representations of reality that can be used to explore the mechanisms that drive complex phenomena that other statistical methods would struggle to accommodate (Gilbert and Troitzsch, 2005; Rauch, 2002). Popular simulation methods include cellular automata, agent-based models, artificial neural networks and genetic algorithms (Brownlee, 2007). The process of model-building has been described as a 'third way' of conducting research because it combines deductive and inductive approaches into a single pragmatic framework (Axelrod, 1997).

In simulation methods, a theory or set of theories is formulated to explain a phenomenon, and is then encapsulated into a structured programming language and simulation software (e.g. Netlogo, Repast, C++). The model is then analysed to see if it can generate the phenomenon of interest, and theories can be modified based upon model outputs. Outputs from the model can also generate new research questions, while additional empirical data can be integrated into the model. This process is illustrated in Figure 10.5. The rigorous iterative process of formalizing theories, model-building, model analysis and retesting can help researchers to better understand complex systems at a much deeper level than by using statistical methods alone (Gilbert and Troitzsch, 2005).

To give an example of these concepts, Figure 10.6 illustrates an 'ecological theory' of poaching using agent-based modelling techniques. Agent-based modelling (ABM) is one type of simulation technique that

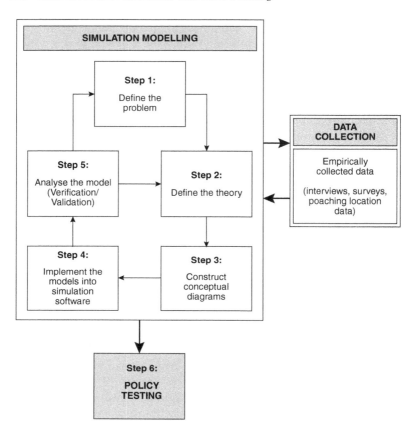

Figure 10.5   The model building process. Simulation outputs can generate new research questions and gaps for data collection, which in turn can be integrated back into the model

can explore how the behaviours and decisions of interacting individuals or agents within their environment can generate complex macrolevel patterns (Gilbert and Troitzsch, 2005). It is particularly appropriate for studying complex systems and has been utilized to investigate animal foraging behaviour (Grimm and Railsback, 2005), patterns of crime (Malleson et al., 2010) and wildlife hunting (Hill et al., 2014; Ling and Milner-Gulland, 2008).

The agents in the model in Figure 10.6 are identified (animals, poachers and rangers) with their objectives (e.g. an animal agent's objective is to forage for food). Each component of this wildlife poaching

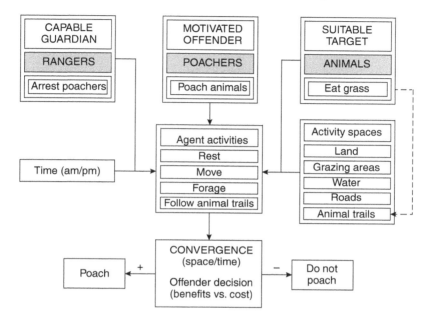

Figure 10.6 A conceptual diagram illustrating an ecological model of poaching. Agents are shown, together with how their goals and their behavioural activities within possible activity spaces may or may not lead to a poaching event

system draws from opportunity perspectives of crime and poaching behaviour in the literature. For example, activity spaces (Brantingham and Brantingham, 1981) can represent grazing areas, water points or animal trails, which in turn can influence the type of behaviour that agents may engage in to meet their objectives. The model also captures Lemieux's (2014: 6) 'triple foraging process', where animals forage for food, poachers forage for animals and rangers forage for poachers, which determines poaching opportunities. How these actors converge at a particular space and time will influence a person's decision on whether to poach based on the perceived costs and benefits (Clarke and Cornish, 1985).

More details about the ecological theory of poaching can be found in the work of Hill and colleagues (2014: 120), but the underlying premise is that, similar to urban crimes such as burglary, poaching clusters in space-time 'poaching hotspots' as a result of the particular routine activity and foraging behaviours of rangers, poachers and animals. This ecological theory of poaching was used as a foundation to construct their ABM of poaching for an area of Queen Elizabeth National Park in Uganda, whose aim was to better understand why rangers observe only a small fraction of poaching activities when they patrol. Figure 10.7 provides a visual layout of the model using the Netlogo ABM software package (Wilensky, 1999). The highly visual nature of ABM software is one of the appeals and strengths of verifying and validating models because it allows a researcher to observe the behaviours of individual agents and how a phenomenon might emerge over (simulated) time as a result of their interactions. Moreover, simulation models might help to direct scarce empirical research resources to where they can make the most useful discoveries. For example, the lack of data on poaching behaviour that was problematic for the models that were developed by Hill and colleagues (2014) demonstrates the need to undertake

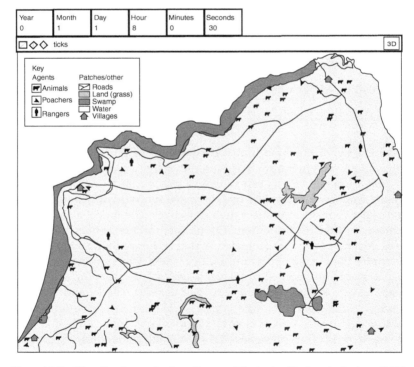

Figure 10.7 Visualization of the Queen Elizabeth National Park wildlife poaching Netlogo model showing environmental features and mobile agents. The six monitors in the top-left corner represent elapsed simulated time

additional interview-based research to better understand poaching practices and the motivations of those people who poach. Such research
might also identify possible alternatives to poaching from the people
who are committing this type of offence, which in turn can then be
tested in future models.

Yet possibly the most advantageous reason for using simulation models over other statistical and ethnographic methods is that they allow
researchers to ask 'what if'-type questions and to test the outcomes
of different policies that one might not be able to achieve in reality for financial, logistical or ethical reasons (Hughes et al., 2012). For
example, if a park manager has limited financial resources to spend
on anti-poaching strategies, they might want to test the effectiveness
of increasing the number of rangers or to extending the time spent
patrolling. Figure 10.8 shows the number of poacher agents who were
detected by ranger agents when the Queen Elizabeth National Park
model was run for 18 simulated months to test four different patrol
strategies (time spent patrolling versus number of patrols). If we assume
that the financial cost is the same for each strategy, the models suggest
that increasing the number of hours spent patrolling but with fewer
rangers may be more effective at detecting poaching than increasing
the number of rangers with a reduction in time patrolling. This raises a
further interesting question: Is this a realistic output as a result of these

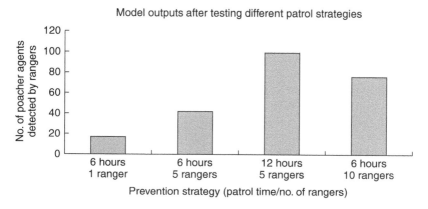

Figure 10.8 A simple graph showing different poaching detection rates by ranger
agents in the Queen Elizabeth National Park simulation model (over 18 simulated
months) when varying the number of hours patrolling versus varying the number
of ranger agents

particular biological-social interactions, or simply the result of the particular construction and parameterization of the model? This question can then be explored using ethnographic and interview methods or further model testing.

Although the example given in Figure 10.8 illustrates a pro-enforcement poaching-reduction strategy, the models can be adapted to test the effectiveness of pro-community methods. For example, one could compare patrol strategies with providing communities with a certain amount of meat or money. The models can also be run over several simulated years within the space of hours or days on a desktop computer. This ability to test the effectiveness of different policy strategies over extended periods of (simulated) time could prove especially useful in situations where financial resources (and time) are limited.

Discussion and future work

Recently von Essen and colleagues (2014) argued that overemphasis on individual behaviour (or normative explanations) is too reductionist to allow an understanding of a complex phenomenon such as poaching. In contrast, systems thinking and complexity theories are by their nature anti-reductionist and demonstrate how complex systems cannot be understood by looking at their individual component parts. Instead it is the interconnectedness and interactions between individuals and the environment that can lead to the emergence of complex behaviour. Some notable examples include Epstein and Axell's (1996) sugarscape models that show how the emergence of violent offenders and genocide can occur as result of the perception of police force numbers at the level of the individual agent (see Epstein, 2002; Rauch, 2002). As a second example, Schelling's (1971) segregation model shows how minor preferences of individual households to live with similar ethnic backgrounds can rapidly lead to the racial segregation that is observed in many urban societies. Nevertheless, it is clear that both micro- and macrolevel mechanisms have their role to play in understanding all kinds of complex social phenomena, and this is no doubt the case for poaching as well. Therefore, taking a systems thinking perspective, I argue that poaching can be considered as a complex biological-social system which is driven by particular political, social, environmental and individual interactions, and from which aggregate patterns of poaching hotspots emerge. Whether the observed patterns can be fully accounted for by instrumental or normative theories, or whether other theoretical elements need to supplement them, at the level of individual events/actions or at higher emergent levels, is for simulation and real-world research to determine.

Nevertheless, careful consideration is required regarding the construction of the models, particularly in determining the level of necessary complexity (e.g. which agents and variables are important to describe the system). This requires considerable time, patience and practice of learning how to programme, build and analyse simulation models. Thankfully, excellent publications and guides are available (e.g. Axelrod and Tesfatsion, 2006; Railsback and Grimm, 2011), and online help forums can be used to discuss and check coding.[7] It is also essential that researchers take the time to verify and validate the models of interest. Verification ensures that the model has been programmed correctly and that it behaves as expected, thereby increasing its internal validity (Townsley and Johnson, 2008). Validation refers to the processes of examining whether the model is able to represent the system that it is attempting to simulate (Casti, 1997). This is a crucial step in the model-building process because any conclusions drawn from the model outputs rely on the accurate representation of the system under study. Indeed, if modelling is to be accepted as a scientific approach then models must be falsifiable (Malak and Paredis, 2004; Popper, 1972), and researchers need to justify which models (hypotheses) are more likely to be effective when similar outputs are generated by different combinations of variables and interactions. Moreover, outputs from an empirically validated model may be more likely to be seen as credible by policy-makers (Midgley et al., 2007), particularly if any suggested strategies appear to be counterintuitive or against standard practice. In the case of poaching, empirical data about aggregate patterns of poaching, such as poaching events collected by rangers as they patrol and interviews of communities and rangers, can be used to test and validate different models and to measure the effectiveness of different poaching-reduction strategies.

In terms of future research, one of the first steps will be to expand upon the work by Hill and colleagues (2014) to begin building different kinds of simulation model to test instrumental and normative mechanisms and policy strategies as a result of different kinds of poaching motivation. Second, interviews with communities and people who poach will be undertaken to better explore their motivations and hunting strategies. This research might also help to identify possible ways to curb poaching with input from the very people who are committing the offence – information which can then be incorporated into the models.

Finally, somewhere down the road this research could be expanded by exploring whether principles from critical systems heuristics can be integrated with offender motivational-situational frameworks, such as the conjunction of criminal opportunity (Ekblom, 2010). Such a holistic approach might enable a more inclusive dialogue between communities, rangers and conservationists to develop more ethical and holistic solutions to poaching.

Conclusion

Poaching is a complex biological-social system that is driven by normative and instrumental mechanisms that are dependent on the different motivations of people who commit the offence. Looking at the poaching problem through the lens of systems thinking gives a researcher the ability to construct and test theories to explain poaching at a much more holistic, systematic and deeper level than instrumental or normative explanations can alone. It is hoped, therefore, that this research will add to the fascinating theoretical debates about poaching prevention, as well as encourage the development of fair and empirically driven poaching-reduction solutions. Perhaps it is best to conclude by re-emphasizing the quotation by Epstein and Axtell (1996: 20) at the beginning of this chapter and by encouraging all of those who seek to better understand and prevent environmental crime or to challenge the dominant power structures to ask themselves the following question: 'Can I grow it?'

Notes

1. Unless otherwise stated, 'poaching' refers to the illegal taking of plants or animals.
2. The reader is referred to Sollund (2011: 438) for an alternative animal-rights critique of the terms 'wildlife' and 'poaching'.
3. Many aspects of poaching are essentially the same as hunting (both may use snares and pursue animals), and the two largely differ only insofar as hunting is illegal in some jurisdictions.
4. For example, at the time of writing, the well-known Kruger National Park in South Africa is approximately 20,000 km^2 – roughly equivalent to the size of Israel or Wales.
5. The recognition of the different motivations for poaching may also explain, in part, the growing unease among researchers about defining someone as 'a poacher' because of its negative connotations. One solution is to disaggregate the act from the person, similar to how one might describe someone who suffers from a medical condition (e.g. 'a person with autism' rather than 'an autistic person'). Therefore this chapter mainly refers to 'a person who poaches' or 'a person that poached'. The term 'poacher' will only be used

when discussing a person's behaviour during the act of poaching because, arguably, this is their main identity during this activity.
6. Crime opportunity theory is equivalent to the more commonly used term of 'environmental criminology' (Wortley and Mazerolle, 2008). The former term is utilized throughout the chapter in order to reduce any confusion with the field of 'environmental crime', which relates to the research of crimes that harm the environment (White, 2013, p. 4).
7. See OpenABM (www.openabm.org.uk) and other Yahoo groups (https:// groups.yahoo.com/neo/groups/netlogo-users/info).

Bibliography

Arnold, T.J.R. (1998) *Introduction to Materials Management*, 3rd edn. Englewood Cliffs, NJ: Prentice Hall.

Axelrod, R.M. (1997) *The Complexity of Cooperation: Agent-Based Models of Competition and Collaboration.* New Jersey: Princeton University Press.

Axelrod, R. and Tesfatsion, L. (2006) 'Appendix AA guide for newcomers to agent-based modeling in the social sciences', *Handbook of Computational Economics*, 2: 1647–1659.

Ayling, J. (2013) 'What sustains wildlife crime? Rhino horn trading and the resilience of criminal networks', *Journal of International Wildlife Law & Policy*, 16(1): 57–80.

Barnaud, C., Le Page, C., Dumrongrojwatthana, P. and Trébuil, G. (2010) Exploring synergies between farmers' livelihoods, forest conservation and social equity participatory simulations for creative negotiation in Thailand highlands. Communication with oral presentation at the ISDA conference (Innovation and Sustainable Development for Agriculture and Food), Montpellier, France, 28 June–1 July 2010.

Becker, G.S. (1968) 'Crime and punishment: An economic approach', *The Journal of Political Economy*, 76: 169–217.

Bersini, H. (2012) 'Emergent phenomena belong only to biology', *Synthese*, 185: 257–272.

Borrion, H. (2013) 'Quality assurance in crime scripting', *Crime Science*, 2(1): 6.

Bouché, P., Renaud, P.C., Lejeune, P., Vermeulen, C., Froment, J.M., Bangara, A. and Fay, M. (2010) 'Has the final countdown to wildlife extinction in Northern Central African Republic begun?' *African Journal of Ecology*, 48(4): 994–1003.

Brantingham, Patricia and Brantingham, Paul (1995) 'Criminality of place', *European Journal on Criminal Policy and Research*, 3(3): 5–26.

Brantingham, P.L. and Brantingham, P.J. (1981) 'Notes of the geometry of crime'. In Brantingham, P.J. and Brantingham, P.L. (ed.) *Environmental Criminology*. Prospect Heights, IL: Waveland Press, pp. 27–54.

Brownlee, J. (2007) Complex adaptive systems. Complex intelligent systems laboratory, Technical Report 070302A, Swinburne University of Technology, Melbourne, Australia.

Büscher, B., Sullivan, S., Neves, K., Igoe, J. and Brockington, D. (2012) 'Towards a synthesized critique of neoliberal biodiversity conservation', *Capitalism Nature Socialism*, 23(2): 4–30.

Cabrera, D., Colosi, L. and Lobdell, C. (2008) 'Systems thinking', *Evaluation and Program Planning,* 31: 299–310.

Casti, J. (1997) *Would-Be-Worlds: How Simulation is Changing the Frontiers of Science.* New York: John Wiley and Sons Ltd.

Chainey, S., Tompson, L. and Uhlig, S. (2008) 'The utility of hotspot mapping for predicting spatial patterns of crime', *Security Journal,* 21(1): 4–28.

Checkland (1981) *Systems Thinking, Systems Practice,* Chichester: Wiley.

Clarke, R.V. and Cornish, D.B. (1985) 'Modeling offenders' decisions: A framework for research and policy', *Crime and Justice,* 6: 147.

Clarke, R.V. and Eck, J.E. (2005) Crime Analysis for Problem Solvers. Washington, D.C.: Center for Problem Oriented Policing.

Coad, L. (2007) Bushmeat hunting in Gabon: Socio-economic and hunter behaviour. PhD thesis, University of Cambridge.

Coad, L., Abernethy, K., Balmford, A., Manica, A., Airey, L. and Milner-Gulland, E.J. (2010) 'Distribution and use of income from bushmeat in a rural village, central Gabon', *Conservation Biology,* 24(6): 1510–1518.

Cochrane, M.A. (2003) 'Fire science for rainforests', *Nature,* 421: 913–919.

Cohen, L.E. and Felson, M. (1979) 'Social change and crime rate trends: A routine activity approach', *American Sociological Review,* 44: 588–608.

Cornish, D.B. (1994) 'The procedural analysis of offending and its relevance for situational prevention', *Crime Prevention Studies,* 3: 151–196.

Davies, T.E., Wilson, S., Hazarika, N., Chakrabarty, J., Das, D., Hodgson, D.J. and Zimmermann, A. (2011) 'Effectiveness of intervention methods against crop-raiding elephants', *Conservation Letters,* 4(5): 346–354.

Distefano, E. (2005) 'Human-wildlife conflict worldwide: Collection of case studies, analysis of management strategies and good practices', *SARD. Initiative Report,* FAO, Rome.

Dudley, N. (ed.) (2008) *Guidelines for Applying Protected Area Management Categories.* Gland, Switzerland: IUCN.

Dudley, J.P., Ginsberg, J.R., Plumptre, A.J., Hart, J.A. and Campos, L.C. (2002) 'Effects of war and civil strife on wildlife and wildlife habitats', *Conservation Biology,* 16(2): 319–329.

Duffy, R. (2010) *Nature Crime: How We're Getting Conservation Wrong.* New Haven, CT: Yale University Press.

Duffy, R. (2014) 'Waging a war to save biodiversity: The rise of militarized conservation', *International Affairs,* 90(4): 819–834.

Ekblom, P. (2010) *Crime Prevention, Security and Community Safety with the 5Is Framework.* Basingstoke: Palgrave Macmillan.

Ekblom, P. and Tilley, N. (2000) 'Going equipped', *British Journal of Criminology,* 40(3): 376–398.

Eliason, S.L. (1999) 'The illegal taking of wildlife: Toward a theoretical understanding of poaching', *Human Dimensions of Wildlife,* 4(2): 27–39.

Eliason, S.L. (2004) 'Accounts of wildlife law violators: Motivations and rationalizations', *Human Dimensions of Wildlife,* 9: 119–131.

Eliason, S.L. and Dodder, R.A. (1999) 'Techniques of neutralization used by deer poachers in the western United States: A research note', *Deviant Behavior,* 20(3): 233–252.

Enticott, G. (2011) 'Techniques of neutralising wildlife crime in rural England and Wales', *Journal of Rural Studies,* 27(2): 200–208.

Epstein, J. (2002) 'Modeling civil violence: An agent-based computational approach', *Proceedings of the National Academy of Science of the U.S.A*, 99 (Suppl. 3): 7243–7250.

Epstein, J.M. and Axtell, R. (1996) *Growing Artificial Societies: Social Science From the Bottom Up*. Washington, D.C./MA: Brookings Institution Press/MIT Press.

Etzioni, A. (1988) *The Moral Dimension: Toward a New Economics*. New York: Free Press.

Felson, M. (1987) 'Routine activities and crime prevention in the developing metropolis', *Criminology*, 25(4): 911–932.

Filteau, M. (2012) 'Deterring defiance: 'Don't Give a Poacher a Reason to Poach', *Journal of Rural Criminology*, 1: 236–255.

Forsyth, C.J. and Marckese, T.A. (1993) 'Thrills and skills: A sociological analysis of poaching', *Deviant Behavior*, 14: 157–172.

Gallic, B.L. and Cox, A. (2006) An economic analysis of illegal, unreported and unregulated (IUU) fishing: Key drivers and possible solutions. *Marine Policy*, 30(6): 689–695.

Gilbert, G.N. and Troitzsch, K.G. (2005) *Simulation for the Social Scientist*. Maidenhead, England, New York, NY: Open University Press.

Grimm, V. and Railsback, S.F. (2005) *Individual-Based Modeling and Ecology*. New Jersey: Princeton University Press.

Guitart, R., Croubels, S., Caloni, F., Sachana, M., Davanzo, F., Vandenbroucke, V. and Berny, P. (2010) 'Animal poisoning in Europe. Part 1: Farm livestock and poultry', *The Veterinary Journal*, 183(3): 249–254.

Hamilton, M.D. and Erhart, E.M. (2012) 'Forensic evidence collection and cultural motives for animal harvesting'. In Huffman, J.E. and Wallace, J.R. (eds) *Wildlife Forensics: Methods and Applications*. Hoboken: John Wiley & Sons Ltd, pp. 65–76.

Harrison, R.D. (2011) 'Emptying the forest: Hunting and the extirpation of wildlife from tropical nature reserves', *BioScience*, 61: 919–924.

Heckenberg, D. and White, R. (2013) 'Innovative approaches to researching environmental crime'. In South, N. and Brisman, A. (eds) *Routledge International Handbook of Green Criminology*. Milton Park, Oxon: Routledge, pp. 85–103.

Hilborn, R., Arcese, P., Borner, M., Hando, J., Hopcraft, G., Loibooki, M. and Sinclair, A.R. (2006) 'Effective enforcement in a conservation area', *Science*, 314(5803): 1266–1266.

Hill, J.F., Johnson, S.D. and Borrion, H. (2014) 'Potential uses of computer agent-based simulation modelling in the evaluation of wildlife poaching'. In Lemieux, A. (ed.) *Situational Prevention of Poaching*. New York: Routledge.

Hitchcock, R.K. (2000) 'Traditional African wildlife utilization: Subsistence hunting, poaching, and sustainable use'. In Prins, H.H.T., Grootenhuis, J.G. and Dolan, T.T. (eds) *Wildlife Conservation by Sustainable Use*. Netherlands: Springer, pp. 389–415.

Holland, J. (1992) 'Complex adaptive systems', *Daedalus*, 121(1): 17–30.

Holmes, G. (2007) 'Protection, politics and protest: Understanding resistance to conservation', *Conservation and Society*, 5(2): 184.

Hughes, H.P., Clegg, C.W., Robinson, M.A. and Crowder, R.M. (2012) 'Agent-based modelling and simulation: The potential contribution to organizational psychology', *Journal of Occupational and Organizational Psychology*, 85(3): 487–502.

Jackson, M.C. (2003) *Systems Thinking: Creative Holism for Managers*. Chichester, UK: John Wiley & Sons Ltd.

Johnson, S.D. (2010) 'A brief history of the analysis of crime concentration', *European Journal of Applied Mathematics*, 21(4–5): 349–370.

Kahler, J.S. and Gore, M.L. (2012) 'Beyond the cooking pot and pocket book: Factors influencing noncompliance with wildlife poaching rules', *International Journal of Comparative and Applied Criminal Justice*, 36(2): 103–120.

Kernick, D. (2004) *Complexity and Healthcare Organization: A View From the Street*. Abingdon, UK: Radcliffe Medical Press.

Kideghesho, J.R. (2009) 'The potentials of traditional African cultural practices in mitigating overexploitation of wildlife species and habitat loss: Experience of Tanzania', *International Journal of Biodiversity Science & Management*, 5(2): 83–94.

Klockars C.B. (1974) *The Professional Fence*. New York: Macmillan.

Kroneberg, C., Heintze, I. and Mehlkop, G. (2010) 'The interplay of moral norms and instrumental incentives in crime causation', *Criminology*, 48(1): 259–294.

Kumar, S.A. and Suresh, N. (2006) *Production and Operations Management*. New Delhi: New Age International.

Lahm, S. (2001) 'Hunting and wildlife in Northeastern Gabon. why conservation should extend beyond protected areas'. In Weber, W., White, L.J.T. and Vedder, A. (eds) *African Rain Forest Ecology and Conservation: An Inter Disciplinary Perspective*. New York, NY: Yale University Press, pp. 344–354.

Larsen, K.R. (2011) 'Critical systems thinking for the facilitation of conservation planning in Philippine coastal management', *Systems Research and Behavioral Science*, 28(1): 63–76.

Lemieux, A.M. (ed.) (2014) *Situational Prevention of Poaching*. New York: Routledge.

Lemieux, A.M., Bernasco, W., Rwetsiba, A., Guma, N., Driciru, M. and Kirya, H.K. (2014) 'Tracking poachers in Uganda', *Situational Prevention of Poaching*. New York: Routledge.

Lindsey, P., Balme, G., Becker, M., Begg, C., Bento, C., Bocchino, C., Dickman, A., Diggle, R., Eves, H., Fearnhead, P., Henschel, P., Lewis, D., Marnewick, K., Mattheus, J., McNutt, J.W., McRobb, R., Midlane, N., Milanzi, J., Morley, R., Murphree, M., Nyoni, P., Opyene, V., Phadima, J., Purchase, N., Rentsch, D., Roche, C., Shaw, J., van der Westhuiz en, H., Van Vliet, N. and Zisadza, P. (2012) *Illegal Hunting and the Bushmeat Trade in Savannah Africa: Drivers, Impacts and Solutions to Address the Problem*. New York: Panthera/Zoological Society of London/Wildlife Conservation Society report.

Ling, S. and Milner-Gulland, E.J. (2008) 'Developing an artificial ecology for use as a strategic management tool: A case study of ibex hunting in the North Tien Shan', *Ecological Modelling*, 210: 15–36.

Loibooki, M., Hofer, H., Campbell, K.L. and East, M.L. (2002) 'Bushmeat hunting by communities adjacent to the Serengeti National Park, Tanzania: The importance of livestock ownership and alternative sources of protein and income', *Environmental Conservation*, 29(03): 391–398.

Malak, R.J. and Paredis, C.J. (2004) 'On characterizing and assessing the validity of behavioral models and their predictions', *ASME Design Technical Conferences, DETC2004-57452*, Salt Lake City, Utah, 28 September—2 October 2004, pp. 325–336.

Malleson, N., Heppenstall, A. and See, L. (2010) 'Crime reduction through simulation: An agent-based model of burglary', *Computers, Environment and Urban Systems*, 34: 236–250.

Martín, A.M., Salazar-Laplace, M.E. and Ruiz, C. (2008) 'The sequential analysis of transgressors' accounts of breaking environmental laws', *The Spanish Journal of Psychology*, 11(01): 115–124.

Mattessich, R. (1982) 'The systems approach: Its variety of aspects', *Journal of the Association for Information Science and Technology*, 33: 383–394.

Mcdonald, R.I., Forman, R.T., Kareiva, P., Neugarten, R., Salzer, D. and Fisher, J. (2009) 'Urban effects, distance, and protected areas in an urbanizing world', *Landscape and Urban Planning*, 93(1): 63–75.

McCorquodale, S.M. (1997) 'Cultural contexts of recreational hunting and native subsistence and ceremonial hunting: Their significance for wildlife management', *Wildlife Society Bulletin*, 25(2): 568–573.

Meadows, D.H. (2008) *Thinking in Systems: A Primer*. Chelsea: Green Publishing.

Mendelson, S., Cowlishaw, G. and Rowcliffe, J.M. (2003) 'Anatomy of a bushmeat commodity chain in Takoradi, Ghana', *Journal of Peasant Studies*, 31(1): 73–100.

Messer, K.D. (2010) 'Protecting endangered species: When are shoot-on-sight policies the only viable option to stop poaching?' *Ecological Economics*, 69(12): 2334–2340.

Midgley, G. (2006). Systems thinking for evaluation. In: Williams, B. and Imam, I. *Systems Concepts in Evaluation: An Expert Anthology*. Point Reyes, CA: Edge Press, pp. 11–34.

Midgley, D., Marks, R. and Kunchamwar, D. (2007) 'Building and assurance of agent-based models: An example and challenge to the field', *Journal of Business Research*, 60: 884–893.

Mingers, J. and White, L. (2010) 'A review of the recent contribution of systems thinking to operational research and management science', *European Journal of Operational Research*, 207: 1147–1161.

Mitchener-Nissen, T., Bowers, K. and Chetty, K. (2012) 'Public attitudes to airport security: The case of whole body scanners', *Security Journal*, 25: 229–243.

Moore, L. (2011) 'The neoliberal elephant: Exploring the impacts of the trade ban in ivory on the commodification and neoliberalisation of elephants', *Geoforum*, 42(1): 51–60.

Moreto, W.D. and Clarke, R.V. (2011) *Reasoning Poachers: A General Typology. Presented at the 20th Environmental Criminology and Crime Analysis Symposium.* South Africa: Durban.

Musgrave, R.S., Parker, S. and Wolok, M. (1993) 'Status of poaching in the United States-are we protecting our wildlife', *The Natural Resources Journal*, 33, 977–1014.

Muth, R.M. and Bowe Jr, J.F. (1998) 'Illegal harvest of renewable natural resources in North America: Toward a typology of the motivations for poaching', *Society & Natural Resources*, 11(1): 9–24.

Neumann, R.P. (2004) 'Moral and discursive geographies in the war for biodiversity in Africa', *Political Geography*, 23(7): 813–837.

Noss, A.J. (1998) 'The impacts of cable snare hunting on wildlife populations in the forests of the central African Republic, *Conservation Biology*, 12: 390–398.

Nurse, A. (2013) 'Perspectives on criminality in wildlife'. In Walter, R., Solomon Westerhuis, D. and Wyatt, T. (eds) *Emerging Issues in Green Criminology: Exploring Power, Justice and Harm*. Basingstoke: Palgrave Macmillan.

Okello, M.M., Wishitemi, B.E. and Lagat, B. (2005) 'Tourism potential and achievement of protected areas in Kenya: criteria and prioritization', *Tourism Analysis*, 10(2): 151–164.

Olivier, R.C.D. and Laurie, W.A. (1974) 'Habitat utilization by hippopotamus in the Mara River', *African Journal of Ecology*, 12: 249–271.

Olupot, W., McNeilage, A.J. and Plumptre, A.J. (2009) *An Analysis of Socioeconomics of Bushmeat Hunting at Major Hunting Sites in Uganda*. Working paper no. 38. New York: Wildlife Conservation Society.

Pires, S.F. (2012) 'The illegal parrot trade: A literature review', *Global Crime*, 13(3): 176–190.

Popper, K.R. (1972 [1934]) *The Logic of Scientific Discovery*. London: Hutchinson.

Railsback, S.F. and Grimm, V. (2011) *Agent-Based and Individual-Based Modeling: A Practical Introduction*. Princeton University Press.

Rauch, J. (2002) 'Seeing around corners', *The Atlantic Monthly*, 289(4): 35–48.

Saj, T.L., Mather, C. and Sicotte, P. (2006) 'Traditional taboos in biological conservation: The case of colobus vellerosus at the Boabeng-Fiema monkey sanctuary, Central Ghana', *Social Science Information*, 45(2): 285–310.

Sattler, S., Graeff, P. and Willen, S. (2013) 'Explaining the decision to plagiarize: An empirical test of the interplay between rationality, norms, and opportunity', *Deviant Behavior*, 34(6): 444–463.

Schelling, T.C. (1971) 'Dynamic models of segregation', *Journal of Mathematical Sociology*, 1: 143–186.

Simon, H.A. (1962) 'The architecture of complexity', *Proceedings of the American Philosophical Society*, 106(6): 467–482.

Simon, H.A. (1990) 'Invariants of human behavior', *Annual Review of Psychology*, 41(1): 1–20.

Sollund, R. (2011) 'Expressions of speciesism: The effects of keeping companion animals on animal abuse, animal trafficking and species decline', *Crime, Law and Social Change*, 55(5), 437–451.

Sollund, R. (2015) 'With or without a license to kill: Human-predator conflicts and theriocide in Norway'. In Brisman, A., South, N. and White, R. (eds) *Environmental Crime and Social Conflict: Contemporary and Emerging Issues*. Surrey, UK: Ashgate.

South, N. and Wyatt, T. (2011) 'Comparing illicit trades in wildlife and drugs: An exploratory study', *Deviant Behavior*, 32(6): 538–561.

Steinmetz, R., Chutipong, W., Seuaturien, N., Chirngsaard, E. and Khaengkhetkarn, M. (2010) 'Population recovery patterns of Southeast Asian ungulates after poaching', *Biological Conservation*, 143(1): 42–51.

Sutherland, E.H. (1947) *Principles of Criminology*, 4th ed. Philadelphia, PA: Lippincott.

Swift, L., Hunter, P.R., Less, A.C. and Bell, D.J. (2007) 'Wildlife trade and the emergence of infectious diseases', *EcoHealth*, 4: 25–30.

Sykes, Gresham M. (1978) *Criminology*, New York: Harcourt Brace Jovanovich.

Sykes, G.M. and Matza, D. (1957) 'Techniques of neutralization: A theory of delinquency', *American Sociological Review*, 22: 664–670.

Tacconi, L. (ed.) (2007) *Illegal Logging: Law Enforcement, Livelihoods and the Timber Trade*. London: Earthscan.

Thorngate, W. and Tavakoli, M. (2009) 'Simulation, rhetoric, and policy making'. *Simulation and Gaming*, 40: 513–527.

Townsley, M. and Johnson, S. (2008) 'The need for systematic replication and tests of validity in simulation', In Liu, L. and Eck, J. (eds) *Artificial Crime Analysis Systems*. Information Science Reference, pp. 1–18.

Ulrich, W. (1983) *Critical Heuristics of Social Planning: A New Approach to Practical Philosophy*. Chichester, UK: John Wiley & Sons.

von Bertalanffy, L. (1950) 'An outline of general system theory', *British Journal for the Philosophy of Science*, 1(2): 134–165.

von Essen, E., Hansen, H.P., Källström, H.N., Peterson, M.N. and Peterson, T.R. (2014) 'Deconstructing the poaching phenomenon: A review of typologies for understanding illegal hunting', *British Journal of Criminology*, 54(4): 632–651.

Wall, T. and McClanahan, B. (2015) 'Weaponizing conservation in the 'Heart of Darkness': The war on poachers and the neocolonial hunt'. In Brisman, A., South, N. and White, R. (eds) *Environmental Crime and Social Conflict: Contemporary and Emerging Issues*. Surrey, UK: Ashgate.

Warchol, G.L. and Johnson, B.R. (2011) 'Securing national resources from theft: An exploratory theoretical analysis', *Journal of Applied Security Research*, 6(3): 273–300.

Ward, N. (1999) 'Foxing the nation: The economic (in) significance of hunting with hounds in Britain', *Journal of Rural Studies*, 15(4): 389–403.

Wato, Y.A., Wahungu, G.M. and Okello, M.M. (2006) 'Correlates of wildlife snaring patterns in Tsavo West National Park, Kenya', *Biological Conservation*, 132: 500–509.

Watson, F., Becker, M.S., McRobb, R. and Kanyembo, B. (2013) 'Spatial patterns of wire-snare poaching: Implications for community conservation in buffer zones around National Parks', *Biological Conservation*, 168: 1–9.

Weaver, W. (1948) 'Science and complexity', *American Scientist*, 36: 536–544.

Wiener, N. (1948) *Cybernetics*. Cambridge, MA: MIT Press.

Wiens, J.A., Stenseth, N.C., Horne, B.V. and Ims, R.A. (1993) 'Ecological mechanisms and landscape ecology', *Oikos*, 66: 369–380.

Wikström, P.-O.H. (2006) 'Individuals, settings, and acts of crime: Situational mechanisms and the explanation of crime', In Wikström, P.-O.H. and Sampson, R.J. (eds) *The Explanation of Crime. Context, Mechanisms and Development*. Cambridge, UK: Cambridge University Press.

Wilensky, U. (1999) NetLogo, http://ccl.northwestern.edu/netlogo/. Center for Connected Learning and Computer-Based Modeling, Northwestern University, Evanston, IL.

Williams, K. (2012) 'Human needs and norms: Some concepts for exploring sustainable human – environment relationships'. In Bender, H. (ed.) *Reshaping Environments: An Interdisciplinary Approach to Sustainability in a Complex World*. United Kingdom: Cambridge University Press, pp. 335–351.

Woods, M., Anders, J. and Guilbert, S. (2012) ' "The Country(side) Is Angry": Emotion and explanation in protest mobilization', *Journal of Social & Cultural Geography*, 13: 567–585.

Wortley, R. and Mazerolle, L. (2008) *Environmental Criminology and Crime Analysis*. London: Routledge.

11

'Now You See Me, Now You Don't': About the Selective Permissiveness of Synoptic Exposure and Its Impact

Andrea Beckmann

Introduction

This chapter focuses on the Animal Defenders International (ADI) TV advertisement (http://www.youtube.com/watch?v=N6JFb5RHrs8) that was made in order to raise public awareness of the problematic consequences that are entailed in the 'normalized' and commercialized representation, use and abuse (not least the trafficking) of primates (see also Sollund and Wyatt in Westerhuis et al., 2013) for the 'entertainment' of humans. The advert was produced in 2005 as part of a campaign called 'My Mate's a Primate', which attempted to problematize the use of primates in four main areas: entertainment (circuses, films, TV programmes and advertising), the pet trade, experiments and bush meat. While this advert specifically focuses on the trafficking and cruel confinement of primates, its principal and powerful message is a broader one as it makes direct links between the experiences of a non-human and a human animal.

The ADI advert begins by showing a still of the face of a chimpanzee. When the video clip begins, a sound of desperate crying can be heard and the image of a cage in which a young child sits on the floor comes into focus. The child is rocking itself backwards and forwards and when the camera/human moves towards the cage, the child moves away to the other corner of the cage as if in fear. The child then clutches the iron bars of the cage and puts its fingers into its mouth nervously. A voiceover states: 'A chimp has the mental and emotional age of a four-year-old child.' Then a chimpanzee is shown in the same position in which the child was previously, also sucking its fingers nervously. The voiceover continues: 'Although they share 98% of our genetic makeup they are still caged and abused to entertain us. Please make a donation...'.

While human animals are routinely exposed to an anthropocentric perspective and *Weltanschauung*, this advert exposes human animals to a non-human animal phenomenality that Aaltola (2012) called 'zoocentric'. In opposition to the normalized and legitimated instrumental view that sees non-human animals in a merely instrumental fashion, this advert opens the viewer up to a different understanding. This is an empathetic relational understanding that appreciates the shared phenomenal consciousness of animals, both human and non-human.

The advert tried to heighten public awareness of the problematic, painful and ultimately harmful consequences of the 'normalized' and commercialized representation, use and abuse of primates in the 'entertainment' of human animals. On 22 April the Grand Chamber of the European Court of Human Rights held in *Animal Defenders International vs. UK*, by a majority of 9:8, that the UK's broadcasting ban on political advertising under the Communications Act 2003 did not violate the rights to free speech of an animal rights NGO. The NGO – ADI – was refused permission to air the advert.

The ruling of the European Court of Human Rights is the focus of this chapter not only because it contradicts previous Strasbourg case law (e.g. cases concerning Switzerland in 2001 (*VgT Verein gegen Tierfabriken vs. Switzerland*) and Norway in 2008 (*TV Vest As & Rogaland Pensjonistparti vs. Norway*) in which 'essentially identical' prohibitions were ruled to be in violation of Article 10) but also because of the various problematic implications that it clearly has.

The critical criminologist and abolitionist Thomas Mathiesen (1997) developed the concept of 'synopticon' as a supplementation to Foucault's concept of 'panopticon' (1977; based on Bentham's invention of the all-seeing panopticon from 1823): 'such a synoptic apparatus exists symbiotically with the panoptic as a means of generating surveillable cultural enclosures' (Simon, 2005: 10). According to Mathiesen (1997), both processes are endemic to our Western societies. These processes have to be understood as part of what Sollund (2012) refers to as the context of doxic speciesism within consumerist societies.

The total system of modern mass media generates a situation in which masses of people focus on a selected few. This 'synoptic space' performs its visual and continuous power over masses of people through an active process of filtering and shaping the 'informations' (Mathiesen,

1997). The process is, of course, interdependent with a broader context of political and economic interests that are specific to a 'historical field'.

"Synopticon' thus functions in terms of social control through inducing people to specific patterns of self-control and [...] inducing specific patterns of desire [...] which fit the requirements of capitalist consumerism. Although 'lived bodies' are never completely determinable, the effects of the 'synopticon' should not be underestimated." (Beckmann 2001: 72) Simon explained the specific operation of this form of subtle but effective social control as: 'audiences for these media are enculturated rather than trained or disciplined in any formal sense and audience behaviours are structured (though not determined) by the synoptic management of perception, risk, morality, desire and truth'.

(Simon, 2005: 10)

From a critical criminological perspective it is helpful to make use of the concept of 'framing' (e.g. Butler, 2009) here because it allows for the exploration of possible implications of media representations (their selective lack of representation) and their broader consequences as: 'Frames socially construct reality by imposing meaning on actions, events, and issues and organize them within familiar categories and narratives. Following familiar patterns, media select and emphasize certain facts and downplay others; their presentation of issues therefore shapes public perceptions and views of appropriate responses' (Sorenson, 2009: 238). In this specific context, following Butler's work (2009) to include non-human animals, such frames constitute which forms of life (and their suffering) are considered as recognizable and if the loss of specific forms of life are represented as grievable to audiences.

In 'synoptic space' within neoliberal contexts, 'Important issues about politics, power, war, life, and death get either trivialized or excluded from public discourse as a market-driven media culture strives to please its corporate sponsors and attract the audiences it has rendered illiterate' (Giroux, 2008: 164). This UK ban confirmed by a European Court of Human Rights ruling is relevant to critical, as well as green/cultural, criminologists. This chapter will critically explore the interdependencies between the material and ideological 'conditions of domination' as they relate to all 'lived bodies' but especially non-human animals. It is informed by a critical criminological perspective that engages itself with ideology critique (Hess, 1986) as well as aiming to strengthen connections between cultural criminology and green criminology.

The Communications Act 2003: Ensuring an even playing field?

It is obvious that there are strong political reasons that might have influenced the ruling of the European Court of Human Rights if one takes the recent rifts between the UK and the EU into account. Lewis suggested that

> *ADI* might be portrayed as an example of successful dialogue between Strasbourg and the UK's legislature and courts: the former decided to revisit its earlier approach, taking on board the considered views of the latter. A less charitable view is that in the wake of Hirst v UK and the prisoners voting saga, the Strasbourg Court was wary of being accused of again meddling in British democracy and simply lost its nerve, preferring to overturn its previous case law than risk another confrontation with the UK government and press.
>
> (Lewis, 2013)

Apart from such obvious politically opportunist motivations, the European Court of Human Rights' selection of this specific case in order to potentially 'appease' the UK requires further critical exploration as it is the assumed and stated reason for the Communications Act ban to protect democracy in the UK from distortion. This act is meant to prevent wealthy groups from purchasing airtime and clogging the airwaves up with their own political messages.

It is important to acknowledge the problematic interrelationships between corporate power and the media in the context of corporate capitalism, as already pointed out in 1988 by Herman and Chomsky who suggested that market forces mould the performance of the media within the supraframework of state capitalism.

This suggestion is echoed in Edwards and Cromwell's (2006) research findings that established some impressive material that clearly demonstrates that the media's performance is highly influenced by, and biased towards, the interests of established power. Thus the fact that an NGO with an important and predominantly marginalized perspective was prohibited from accessing the same airwaves to which wealthy commercial organizations have unlimited access is clearly unfair and could be interpreted as being part of 'the apparent global criminalisation of animal advocacy...' (Yates, 2011: 469). In this context Sorenson's (2009) observations are specifically useful to take into account, as he notes, 'Mainstream media are large corporations owned by wealthy individuals or other corporations and generally serve the interests of dominant classes. A general pro-capitalist filter operates' (Sorenson, 2009: 238).

Protecting 'democracy', protecting the 'Order of Things'

The ban of the ADI advert has important consequences as the media have a huge impact on the way in which meaning is given to the world and is frequently represented as representing the 'truth'.

The potential scope of this prohibition that was legitimated by the Communications Act 2003 is extremely broad because it encompasses 'social advocacy advertising' and not only political adverts. Any adverts that address matters of 'public controversy' in the UK are included in the proclaimed 'protection' of democracy. This points to an even broader implication of these 'conditions of domination' as

> The façade of modern democracy depends on the idea that we are *already* living in a free and open society-the media are a central plank of this 'necessary illusion'. The maintenance of the deception is vital if elites are to continue manipulating the public to fight wars and to wreck the environment [including non-human animals, note of the author] for profit.
>
> (Edwards and Cromwell, 2006: 201–202)

In this context this chapter understands the ruling of the European Court of Human Rights on 22 April 2013 in favour of the UK government as part of a wider struggle that according to Giroux (2013) originates in neoliberal corporate power, and that in relation to the role of the 'synoptic space' of the media has both symbolic and pedagogical dimensions.

While Giroux's observations focus on the US context, the imposition of neoliberal ideology has had similar consequences in other glocal capitalist contexts. 'Particular sponsors, who purchase advertising, want to maintain a "buying mood" among audiences and their financial power limits content. While information from establishment sources is readily used, information from dissident groups is not... [general pro-capitalist] filters narrow the range of acceptable information...' (Sorenson, 2009: 238).

Various media forms and uses of media time/space are not accessible to alternative interest groups such as ADI due to global corporate interests and power, thus the ban of the advert and the subsequent ruling of the European Court of Human Rights has clear relevance to cultural/green criminology as understood by Brisman and South (2013).

They suggested that 'green criminologists explore those actions that "rupture the normalcy" of everyday life [which this advert clearly does]...' (Brisman and South, 2013: 5). A central part of this 'normalcy of everyday life' is the normality of speciesism (see also Aaltola, 2012; Sollund, 2012) that is embedded in Western modern ideas of 'science', presenting the traditional 'order of things' that appears only to be intensified in late modern times.

Reinforcing Western 'value' systems of order and legitimated violences

As the role of media and communication is clearly emphasized within the context of the 'new global economy' (Virno, 2004), the hierarchization of 'bodies' that is characteristic of the modern Western 'order of things' is likely to be intensified within the 'synoptic space' of media.

To cultural criminology, mediated representations of criminalized behaviour as well as of 'normalized' behaviour are important because they are understood to affect individual and/or collective behaviour. According to Brisman and South (2013), cultural criminology is concerned with the critical exploration of mediated processes of cultural exchange and reproduction as they are a central part of the 'constitution' of people's experience of 'self', 'society' and 'crime' in late modernity.

The selective presence of non-human animals in the media is thus also a matter for cultural criminology's attention as 'cultural criminological attention to images of crime and crime control is also driven by a desire to know what mass media is *not* reporting on – what is *not* being depicted, why such stories are *not* being told, and what are the consequences of these decisions' (Brisman and South, 2013: 9).

In the specific context of this chapter, the ban of the ADI advert excludes audiences from the exposure to imagery that allows for the development of deep empathy with non-human animals. This lack of depiction, of course, continues a long tradition of speciesism that assumes/constructs rigid boundaries, despite the fact that all sentient beings share a common and important similarity: the ability to experience suffering (see also Aaltola, 2012; Sollund, 2012).

Sorenson (2009) states: 'Typically, media coverage of animal rights excludes serious discussion on important issues' (Sorenson, 2009: 240). While concern for non-human animals does have a history in Western traditions of philosophy, genuine empathetic concern for non-human animals has frequently been marginalized and ridiculed (just think of the public response to Friedrich Nietzsche hugging a horse and expressing his empathy for it towards the end of the philosopher's life). In contemporary times the 'terrorism' frame is the predominant frame that is used to represent animal right concerns and the human animals that want to support these (see Sorenson, 2009; Ellefsen et al., 2012; Aaltola, 2012).

The socioculturally and politically constructed and maintained boundaries between humans and non-human animals are, of course, contradictory and, using a post-modern feminist criminological perspective, can be deconstructed in order to expose: 'the problems which reside in the endeavor to keep meaning pure, to say "just this" and not "that", because "just this" always depends on "that" which it is not' (Naffine, 1997: 89).

Western modern epistemology is based on the premise, beginning with Aristotle and 'modernized' by Descartes, of an assumed lack of consciousness that was generalized about all non-human animals, reducing them to instinctual, mechanistic behaviour. Sax suggests that 'modern people have feared above all the "animal" in themselves' (Sax, 2001; Kalof and Fitzgerald, 2007: 276).

It is crucial to keep in mind that at the core of these struggles of meaning lie differing philosophies concerning the relationship between non-human and human animals whereby the traditional rigid anthropocentric philosophy was, and is, privileged at all costs (Aaltola, 2012).

Descartes' 17th-century legacy continues to provide the frame through which Western ideas of philosophy and science, together with contemporary 'understandings' and legitimations in relation to non-human animals, as well as, extending from this, human animal groups that are marginalized, oppressed, exploited and exterminated, are constituted. This is despite the fact that during 'the twentieth century, philosophers made compelling arguments for better treatment of animals drawing analogies with struggles against racism and sexism' (Sorenson, 2009: 241).

While non-human animals, like marginalized and oppressed humans, are frequently absent from media displays, there is at the same time an overexposure of specific representations of non-human animals for humans' entertainment in capitalist consumerist societies, as Sax notes: 'Today anthropomorphic animals are everywhere we look: tigers sell cornflakes and gasoline; talking cows sell milk; bulls and ducks represent sports teams. Centrefolds of undressed women in Playboy are called "bunnies", while in Penthouse, a rival magazine, they are called "pets"...' (Sax, 2001; Kalof and Fitzgerald, 2007: 276). And then, of course, there are endless Disney and Pixar movies, associated corporate products and computerized 'companion' animals that are brimming with representations of non-human animals that are anthropomorphized.

Robotic and virtual pets are, of course, the most technologized of such developments, with problematic consequences. To illustrate the implications of this seemingly selective 'overexposure' to such anthropomorphic representations of non-human animals, Sony's AIBO, a robot dog, is a good example:

> The highest form of virtual pet is one that moves around your room-
> . . . AIBO comes with a number of preprogrammed behavior patterns that encourage owners to project humanlike attributes onto their virtual pets: The AIBO plays, it sleeps, it wags its tail, it simulates feelings of affection and unhappiness. Sony describes the AIBO as 'a true companion with real emotions and instinct' . . . as the technology improves and robot pets become increasingly lifelike, the boundary between people's perceptions of robotic pets and their perceptions of real animals will become increasingly blurred.
>
> (Levy, 2009: 97/98)

Such representations of non-human animals are frequently central aspects of the commodification of leisure, driven by the proliferation and an aggressive marketing of technology. It is crucial not to underestimate the impact that such commodification of 'othered' and colonized life has, as commodities represent the contexts of their production in a distorted and mystifying fashion (e.g. denying the unequal human and non-human social relations that they produce and reproduce (Lefebvre, 1991; Massey, 1994). The continued exploitation of animals and their real 'life worlds' are blended out of the frame of their commercial representation:

> Capitalism *is* animal exploitation and industries such as agribusiness, biotechnology, food, clothing, and pharmaceuticals are major advertisers in media; they do not wish to sponsor a forum for criticism of their operations and products. Thus, institutional structures of media establish a frame that excludes anything other than an instrumental view of animals.
>
> (Sorenson, 2009: 241)

Mills suggests additional reasons for the unacknowledged pledge of non-human animals which he locates in the fact that science framed and thus colonized discourses about animals (Mills; in: Jeffreys, 2013). In a similar vein, Fox Keller explored how one of the sciences that could have opened up humans' empathy and understanding of non-human animals actually ended up shutting down such possibilities as 'In the mid-twentieth century, biology became a "mature science"- that is to say, it succeeded, finally, in breaking through the formidable barrier of "life" that had heretofore precluded it from fully joining the mechanico-reductive tradition of the physical sciences' (Fox Keller, 1992: 113).

This reflects Marcuse's concerns raised in *One-Dimensional Man* (1964) with regard to the intimate connection between 'scientific-technical rationality' and manipulation that he saw as welded together into new forms of social control. To him, technics as a universal form of production is generating an entire *Weltanschauung*, the projection of a historical totality.

To keep this worldview as the dominant one, evidence of cases where so-called 'scientific' research clearly does find parallels in the learnt expressive behaviour of non-human and human animals is not appropriately represented as 'materialist science – in its consistent neo-Darwinist denial of consciousness in any other species apart from humans – cannot bring itself to ascribe what it sees as uniquely human characteristics to non-human animals. Which is ironic, given the imposition of human social constructions…in mainstream presentations of animal behaviour' (Mills; in: Jeffreys, 2013: 3).

The mystification and distortion of shared experiential capacity fits neatly within the ideology of Western supremacy and patriarchy as these also have clearly 'hierarchical implications, and human attempts to define themselves in social terms (e.g. Pasternak, 2007) repeatedly draw on concepts intended to demonstrate human superiority, including language, cooking, intelligence, and curiosity. The human/non-human boundary is, therefore, extremely important, as it serves to justify and uphold paradigms of societal order and progress' (Cassidy and Mills, 2012: 500). This Western modern 'order of things' is crucial to understanding the generation of the hegemony of the present form of rationality (Bradbury-Jones et al., 2007).

Boundaries within Western rationality discourses and in positivist epistemology are not harmless because they ideologically enhance the epistemic authority of the dominant group (McCann and Kim, 2013). The result of such discourses is that dominant groups then will lay claim to neutrality, objectivity and impartiality for themselves.

The inherent violence of modernity's concepts of 'reason', 'rationality' and associated claims of 'civilization' (see also Beckmann, 2009)

that is interdependent of such boundary constructions and their policing was addressed by Chomsky (2013), who elaborated in a recent
'truth-out' article: 'Almost all borders have been imposed and maintained by violence, and are quite arbitrary' (Chomsky, 2013: 1). He
continued:

> The blurring of borders and these challenges to the legitimacy of
> states bring to the fore serious questions about who owns the
> Earth. Who owns the global atmosphere being polluted by the heat-
> trapping gases that have just passed an especially perilous threshold,
> as we learned in may? Or to adopt the phrase used by indigenous
> people throughout much of the world, Who will defend the Earth?
> Who will uphold the rights of nature? Who will adopt the role of
> steward of the commons, ...? That the Earth now desperately needs
> defence from impending environmental catastrophe is surely obvi-
> ous ... At the forefront of the defence of nature are those often called
> 'primitive': members of indigenous and tribal groups, like the First
> Nations in Canada or the Aborigines in Australia – the remnants
> of peoples who have survived the imperial onslaught. At the fore-
> front of the assault on nature are those who call themselves the most
> advanced and civilised: the richest and most powerful nations.

Contextualizing the ban: (The denial of) global suffering

Cassidy and Mills (2012) offer important insights into the socioculturally constructed boundaries between humans, non-human animals and nature. They observe that claims of being human in a 'civilized' society are interdependent with boundary constructions that deem other possibilities of being, and alternative spaces as 'wild', as 'natural'.

The resulting underlying ideology constructs 'nature' and its raw materials (including animals) as mere commodities to which no moral obligation is necessary (Adams, 1990; Lefebvre, 1991). This boundary construction needs to be understood as reflecting larger political/cultural ideologies, such as the normalization of globalized capitalism, heteronormativity, 'gender', 'racialized' power dynamics and so on (see, e.g., Sollund, 2008; Potts and Perry, 2010; Alkon and Agyeman, 2011).

The commodification and objectification of non-human animals is therefore a central plank of the global capitalist, patriarchal and 'racialized' flows of 'production' – the so-called 'free market'. Wrenn (2011) suggests that the global growth of non-human animal use is not a mere reflection and result of the growing affluence of so-called Third World countries but predominantly a consequence of Western domination and manipulation. Dietary choices are reflective of food production and therefore 'non-human animals have become increasingly integral to much of the global economy as part of Western cultural expansion and deliberate food dependency' (Wrenn, 2011: 15).

The UK ban of the ADI advert and the subsequent ruling by the European Court of Human Rights clearly represents an instance of the cultural and political (re)production of the normality of speciesism (Regan, 2003) and of speciesism doxa (Sollund, 2012), the implications of which are illustrated by Regan:

> The bodies of literally billions of animals are intentionally, deliberately, and systematically injured every year, year after year. The freedom of millions of animals is intentionally, deliberately, and systematically denied every minute of every day. The very lives of millions of animals is [sic, are] intentionally, deliberately, and systematically taken every hour of every day.
>
> (Regan, 2003: 119)

Philosophers such as Heidegger and Derrida compared the treatment of animals in agribusiness to the Nazi death camps (Calarco, 2008). This normalized violence has, however, intensified in the present context of neoliberal capitalism and globalization. Wrenn explains:

> When defining and discussing globalisation, we generally do so in reference to human animals. Yet, nonhuman-animals, too, are experiencing much of this phenomenon and are arguably the most impacted by exacerbated inequalities created under neoliberal globalization. Global meat production has increased more than five-fold since 1950 (Nierenberg, 2003). Two-thirds of the increase in meat consumption in 2002 occurred in the developing world (Nierenberg, 2003). In 2009 alone, almost 57 billion non-human animals were slaughtered (Food and Agriculture Organisation of the United Nations, 2009). This figure is not accounting for aquatic non-humans or the exploitation of those not immediately killed for their products.
>
> (Wrenn, 2011: 10)

To Wrenn (2011), the increasing global demand for and reliance on non-human animals as commodities is interdependent with the ideology of capitalism and Western domination.

Conclusion: Troubling the 'naturalness' of human–non-human boundaries

The ban of the ADI advert and its confirmation by the European Court of Human Rights can be interpreted as part of the continuous policing and reinforcing of regulatory institutionalized boundaries that are central to Western capitalist, patriarchal and speciesist 'condition of domination' as 'categories constructed by humans for different *types* of animal, such as domestic/wild; charismatic wildlife/pest; food/non-food. These categories vary across cultures and historical periods, suggesting they must be constantly maintained by societies in order to function in our ordering of human/animal relations (Hytten, 2009; Knight, 2000)' (Cassidy and Mills, 2012: 504).

Currently the rapid development of bioscientific techniques is increasingly challenging the possibility of maintaining the boundary between humans and non-human animals, and bioscientific hybrids are next to impossible to categorize as 'humanized animals' or as 'animalized humans'. While so far limited to the UK, the recent trans-species embryo debate highlighted yet another challenge to the constructed boundary as 'trans-species embryos' were relabelled 'human admixed embryos' (see, e.g., Homer and Davies, 2009).

Other important aspects of contemporary life that concern human and non-human animal relationships are addressed by Sax:

> Today, countless species are driven to extinction, and people, who are becoming increasingly urban, have ever less contact with animals on a daily basis. This has resulted in two seemingly contradictory trends. On the one hand, there is ruthless exploitation of animals in factory farms and in industry. In 1989, an animal was patented for the first time: a mouse genetically engineered to develop cancer. The cloning of a sheep named Dolly in 1997 has largely broken down the boundary between living organisms and manufactured objects. There are now entire varieties called 'pharm animals', developed solely to develop certain chemicals.
>
> (Sax, 2001; Kalof and Fitzgerald, 2007: 276)

Given the many evidently harmful, exploitative and destructive practices that are condoned in the victimization of non-human animals, it seems appropriate to follow Svärd (2008), who suggests that contemporary practices and attitudes towards non-human animals are forms of oppression.

'With the growing popularity of practices policing human/animal, and human/"nature" boundaries, such as biosecurity (e.g. Hinchcliffe and Bingham, 2008), it could be argued that such retributive actions toward animals "out of place" (Douglas, 1966: 36) may be on the rise (Barker, 2010)' (Cassidy and Mills, 2012: 504). In a way, the depiction of and blending in of a human in a context that is designated for non-human suffering in the ADI advert can be seen as a transgressive symbolic act of a human 'out of place' that 'requires' a strong response.

In a context that is characterized by the disappearance of diverse species as well as the waning of global biodiversity, it is important to remind ourselves of the importance of the conditions of and shifts in the visibility of non-human animals that were explored by Berger (1980) some time ago. The visual field of representation, for example, does have a bearing in terms of organizing the way in which non-human animals are seen in a very literal as well as in a metaphorical sense. 'media events come to define the discourses within which a topic circulates . . .' (Cassidy and Mills, 2012: 503). Cassidy and Mills (2012) problematize the lack of focus on media and other forms of communication in the literature on human–non-human animal relations as these are spaces in which such relationships are constituted.

As demonstrated above, contemporary media predominantly display commoditized representations of non-human animals and do not represent their oppressive 'lived realities' in global capitalist and patriarchal contexts, which has severe consequences. 'Under neoliberal capitalism, nonhuman animals are generally understood only as commodities. However, because non-human animals are also sentient, their plight warrants immediate attention' (Wrenn, 2011: 11).

In this context it is helpful to remind ourselves of Jaggar's concept of 'emotional hegemony' (1989). She pointed to the fact that emotionality is not attributed to all bodies equally and thus she refers to the existence of an 'emotional hegemony': 'we absorb the standards and values of our society in the very process of learning the language of emotion, and those standards and values are built into the foundation of our emotional constitution . . . By forming our emotional constitution in particular ways, our society helps to ensure its own perpetuation' (Jaggar, 1989; Kalof and Fitzgerald, 2007: 496). The notion of 'emotional hegemony' (Jaggar, 1989) is not total and some humans experience what Jaggar terms 'outlaw' emotions: emotions that are conventionally unacceptable (e.g. disgust about socially sanctioned modes of relating to and treating 'children' and non-human animals). Representations such as the ADI advert may facilitate the development of such much-needed 'outlaw' emotions and facilitate broader changes as they 'may also enable us to perceive the world differently from its portrayal in conventional descriptions. They may provide the first indications that something is wrong with the way alleged facts have been constructed, with accepted understandings of how things are' (Jaggar, 1989; McCann and Kim, 2013: 497).

In a similar vein but using a framework of moral philosophy, Tom Regan emphasizes:

> If too few of us today are seriously troubled by our contradictory beliefs and attitudes towards animals, I believe this is because too few of us recognize where and why our beliefs and attitudes are contradictory. In particular, too few of us really know what is happening to animals, just as too few of us have ever paused to think carefully about their moral status. What is invisible, both in fact and in value, must first be made visible before it can be seen and understood; contradictions must first be honestly acknowledged before they can be honestly addressed.
>
> (Regan, 2003: 117–118)

The ADI advert did try to do just that via a visual representation of humans' and non-human animals' connectedness, as even so-called 'scientific research' (with all of its 'instrumental fictions') (Nietzsche; in: Angell and Demetis, 2010) has demonstrated many parallels between non-human animals labelled 'chimpanzees' and humans.

'No animal completely lacks humanity, yet no person is ever completely human. By ourselves, we people are simply balls of protoplasm' (Sax, 2001; Kalof and Fitzgerald, 2007: 277). All 'bodies' are 'conjured' by regimes of discourse (Gregg and Seigworth, 2010: 9) that are part of an 'order of things' that demarks a dichotomy between what all sentient beings share – emotion and affect – and so-called 'objective reason'. As is painfully evident, this 'ordering of the world' has and continues to legitimize continuous and disproportionate harms for non-human animals.

It is therefore crucial to engage in the sketching of an alternative philosophy, an ethics, and of pedagogies that facilitate the development of an alternative epistemology that clearly differentiates itself from the dominant challenges that are posed by the anthropocentrism of the Western modern philosophical tradition (Aaltola, 2012). Calarco (2008), following Derrida's initiation of a non-anthropocentric ethics, drew on ethological and evolutionary evidence, and the work of

Heidegger, who called for a radicalized responsibility towards all forms of life. Calarco (2008) suggests adopting a position that acknowledges that while some distinctions are valid, humans and non-human animals are best viewed as part of an ontological whole. He calls for the abolition of classical versions of the human–animal distinction and invites us to devise new ways of thinking about, and living with, non-human animals.

This aim is also reflected in Feenberg's (2002a) work, which contends that a post-technological society has to generate a different kind of science and new types of technology. To him these new dispositions and redesigned technical structures must not be in conflict with the natural environment but in harmony with it. This implies a philosophical reordering, as 'nature' has to be removed from its object position that has been allocated to it within dominant paradigms and become subject. Non-human animals then will finally be treated as subjects in, and with, their own rights.

This ambition to generate an alternative paradigm that turns away from the problematic and intensified ideological legacy of Descartes matches critical criminological concerns as well as concerns of green criminology: 'green criminology must attend to the mediated and political dynamics surrounding the presentation of various environmental phenomena... as well as to resistance to environmental harm and demand for changes in the way that "business as usual" and "ordinary acts of everyday life" (Agnew, 2013) are destroying the environment' (Brisman and South, 2013: 2).

Besthorn (2004, 2013) also suggests that there is a need to facilitate modes of 'active listening', of being enabled to 'hear the voices' of non-human animals which is interdependent with the development of (new/yet unknown) sensitivities. ADI's advert did try to facilitate human animals' ability to develop such sensitivity towards non-human animals.

> compassion and concern for others are central to the best dissident thought... The promise of compassionate dissent is that it provides a powerful, and in fact ever-deepening, motivation- for media activism, peace activism, human and animal rights activism, and environmental activism-in the understanding that compassionate thought and action are also profoundly conducive to our *own* well-being. We need political dissent, but we also need personal, emotional, philosophical – that is fully human-dissent.
>
> (Edwards and Cromwell, 2006: 216–217)

Bibliography

Aaltola, E. (2012) 'Differing philosophies: Criminalisation and the stop Huntingdon animal cruelty debate'. In Ellefsen, R., Sollund, R. and Larsen, G. (eds) *Eco-global Crimes Contemporary Problems and Future Challenges*. Surrey: Ashgate.

Adams, C. (1990) *The Sexual Politics of Meat*. London and New York: Continuum.

Agnew, R. (2013) 'The ordinary acts that contribute to ecocide: A criminological Analysis', *The Routledge International Handbook of Green Criminology*. London: Routledge.

Alkon, A.H. and Agyeman. J. (2011) *Cultivating Food Justice: Race, Class and Sustainability*. Cambridge, MA: MIT Press.

Angell, I.O. and Demetis, D.S. (2010) *Science's First Mistake*. London, New Delhi, New York, Sydney: Bloomsbury.

Barker, K. (2010) 'Biosecure citizenship: Politicising symbiotic associations and the construction of biological threat', *Transactions of the Institute of British Geographers*, 35: 350–363.

Beckmann, A. (2001) 'Deconstructing Myths: The social construction of 'Sadomasochism' versus 'subjugated knowledges' of practioners of consensual 'SM'; in: "Journal of Criminal Justice and Popular Culture" 8(2) pp. 66–95.

Beckmann, A. (2009) *The Social Construction of Sexuality and Perversion: Deconstructing Sadomasochism*. Basingstoke: Palgrave Macmillan.

Beirne, P. (1995) 'The use and abuse of animals in criminology: A brief history and current review', *Social Justice*, 22(1): 5–31.

Beirne, P. (2009) *Confronting Animal Abuse: Law, Criminology, and Human-Animal Relationships'*. Lanham, MD: Rowman and Littlefied.

Berger, J. (1980) *About Looking*. London: Writers and Readers.

Besthorn, F.H. (2004) 'Restorative justice and environmental restoration, twin pillars of a just global environmental policy: Hearing the voice of the victim', *Journal of Societal and Social Policy*, 3(2): 33–48.

Besthorn, F.H. (2013) 'Speaking earth: Environmental restoration and restorative justice'. In van Wormer, K. and Walker, L. (eds) *Restorative Justice Today: Practical Applications*. Los Angeles: SAGE.

Box, S. (1984) *Crime, Power and Mystification*. London: Tavistock.

Bradbury-Jones, C., Sambrook, S. and Irvine, F. (2007) 'Power and empowerment in nursing: A fourth theoretical approach', *Journal of Advanced Nursing*, 62(2): 258–266.

Brisman, A. and South, N. (2013) 'A green-cultural criminology: An exploratory outline', *Crime Media Culture*, published online 6 January 2013, 1–21.

Brisman, A. and South, N. (2014) *Green Cultural Criminology: Constructions of Environmental Harm, Consumerism, and Resistance to Ecocide*. London: Routledge.

Butler, J. (2009) *Frames of War: When Is Life Grievable?* London and New York: Verso.

Calarco, M. (2008) *Zoographies: The Question of the Animal from Heidegger to Derrida*. New York: Columbia University Press.

Cassidy, A. and Mills, B. (2012) ' "Fox Tots Attack Shock": Urban foxes, mass media and boundary-breaching', *Environmental Communication: A Journal of Nature and Culture*, 6(4): 494–511.

Chomsky, N. (2013) 'Noam Chomsky: Who owns the world?' In *Truthout*, Friday 5 July 2013, http://truth-out.org/opinion/item/14373-noam-chomsky-who-owns-the-world.

Douglas, M. (1966) *Purity and Danger: An Analysis of Concepts of Pollution and Taboo*. London: Routledge.

Edwards, D. and Cromwell, D. (2006) *Guardians of Power the Myth of the Liberal Media*. London and Ann Arbor, MI: Pluto Press.

Ellefsen, R., Sollund, R. and Larsen, G. (2012) *Eco-global Crimes Contemporary Problems and Future Challenges*. Surrey: Ashgate.

Feenberg, A. (2002a) *Transforming Technology: A Critical Theory Revisited*. Oxford: Oxford University Press.

Feenberg, A. (2002b) *Heidegger and Marcuse: The Catastrophe and Redemption of History*. New York: Routledge.

Ferrell, J., Hayward, K. and Young, J. (2008) *Cultural Criminology: An Invitation*. London: Sage.

Foucault, M. (1995). *Discipline and Punish: The Birth of the Prison*. New York: Vintage Books.

Fox Keller, E. (1992) *Secrets of Life Secrets of Death*. New York, London: Routledge.

Gregg, M. and Seigworth, G.J. (2010) *The Affect Theory Reader*. Duke University Press.

Gilbert, J. (2008) 'Against the commodification of everything: Anti-consumerist cultural studies in the age of ecological crisis', *Cultural Studies*, 22(5): 551–566.

Giroux, H. (2008) *Against the Terror of Neoliberalism: Politics beyond the Age of Greed*. Boulder, CO: Paradigm Publishers.

Herman, E. and Chomsky, N. (1988) *Manufacturing Consent: The Political Economy of the Mass Media*. New York: Pantheon Press.

Hess, H. (1986)'Kriminalitat als Alltagsmythos. Ein Pladoyer dafur, Kriminologie als Ideologie- kritik zu betreiben', *Kriminologisches Journal*, 1 (Beiheft): 24–44.

Hinchliffe, S. and Bingham, N. (2008) 'Securing life: The emerging practices of biosecurity', *Environment and Planning*, 40(7): 1534–1551.

Homer, H. and Davies, M. (2009) 'The science and ethics of human admixed embryos', *Obstetrics, Gynaecology and Reproductive Medicine*, 19(9): 235–239.

Hytten, K. (2009) 'Dingo dualisms: Exploring the ambiguous identity of Australian Dingoes', *Australian Zoologist*, 35: 18–27.

Jaggar, A. (1989) 'Love and knowledge: Emotion in feminist epistemology'. In McCann, C.R. and Kim, S. (eds) *Feminist Theory Reader*. New York and Oxon: Routledge.

Jeffreys, T. (2013) Representation for animals: An interview with Dr. Brett Mills, *The Journal of Wild Culture*, 23 April 2013, http://www.wildculture.com/article/representation-animals-interview-dr-brett-mills/1142.

Kalof, L. and Fitzgerald, A.J. (2007) *The Animals Reader*. Oxford: BERG.

Kenny, A. (ed./trans.) (1970) *Descartes: Philosophical Letters*. Oxford: Clarendon Press.

Knight, J. (ed.) (2000) *Natural Enemies: People-wildlife Conflicts in Anthropological Perspective* (1–35). London: Routledge.

Lefebre, H. (1991) *The Production of Space*. Oxford: Basil Blackwell.

Lefebvre, H. (1991) *The Critique of Everyday Life*, Vol.1. London and New York: Verso.

Lewis, T. (2013) 'Animal Defenders International vs. UK: A case of fruitful dialogue, or of Strasbourg losing its nerve?', *Oxford Human Rights Hub*, http://ohrh. law.ox.ac.uk/?m=201304.

Marcuse, H. (1964) *One-Dimensional Man*. Boston: Beacon Press.

Massey, D. (1994) *Space, Place and Gender*. Minneapolis: University of Minnesota Press.

Mathiesen, T. (1997) The viewer society: Michel Foucault's 'Panopticon' revisited, *Theoretical Criminology*, 1(2): 215–234.

McCann, C.R. and Kim, S. (2013) *Feminist Theory Reader*, 3rd edn. New York and Oxon: Routledge.

Naffine, N. (1996) *Feminism and Criminology*. Philadelphia: Temple University Press.

Naffine, N. (1997) *Feminism and Criminology*. Cambridge: Polity Press.

Pasternak, C. (2007) *What Makes Us Human?* London: Oneworld Publications.

Potts, A. and Perry, J. (2010) Vegan sexuality, *Feminism & Psychology*, 20(1): 53–72.

Regan, T. (2003) *Animal Rights, HUMAN WRONGS: An Introduction to Moral Philosophy*. New York, Toronto, Oxford: Rowman & Littlefield, Lanham, Boulder.

Sax, B. (2001) 'Animals as tradition'. In Kalof, L. and Fitzgerald, A. (eds) *The Animals Reader*. Oxford, New York: Berg.

Simon, B. (2005) The return of panopticism: Supervision, subjection and the new surveillance, *Surveillance & Society*, 3(1): 1–20.

Sollund, R. (ed.) (2008) *Global Harms Ecological Crime and Speciesism*. New York: Nova Science Publishers.

Sollund, R. (2012) 'Speciesism as doxic practice or evaluating difference and plurality'. In Ellefsen, R. Sollund, R. and Larsen, G. (eds) *Eco-global Crimes. Contemporary Problems and Future Challenges*. Surrey: Ashgate.

Sorenson, J. (2009) Constructing terrorists: Propaganda about animal rights, *Critical Studies on Terrorism*, 2(2): 237–256.

South N. and Brisman A. (eds) (2013) *The Routledge International Handbook of Green Criminology*. London: Routledge.

Svärd, P.-A. (2008) 'Protecting the animals? An abolitionist critique of animal welfarism and green ideology'. In Sollund, R. (ed.) *Global Harms: Ecological Crime and Speciesism*. New York: Nova Science Publishers.

Virno, P. (2004) *A Grammar of the Multitude*. New York: Semiotext(e).

Westerhuis, D., Walters, R. and Wyatt, T. (2013) *Emerging Issues in Green Criminology*. Basingstoke: Palgrave.

Wrenn, C.L. (2011) Resisting the globalization of speciesism: Vegan abolitionism as a site for consumer-based social change, *Journal for Critical Animal Studies* , IX(3): 9.

Yates, R. (2011) 'Criminalising protest about animal abuse. Recent Irish experience in global context', *Crime, Law and Social Change*, 55(5): 469–482.

12

The Occupy Movement vs. Capitalist Realism: Seeking Extraordinary Transformations in Consciousness

Samantha Fletcher

Introduction

In September 2011 the spreading sentiment of 'Ya Basta!' ('Enough!'), stemming from a combination of popular protests within the same decade, particularly those occurring in Spain as part of the Indignados movement (Castañeda, 2012) and from the events commonly identified as the 'Arab Spring', became apparent in the USA. The manifestation was a 'scrappy group of anarchists' (Occupy Wall Street, 2014) who flooded into Liberty Square, New York, to start the movement that is now most commonly known as Occupy Wall Street. The initial call to action came from a blog from Adbusters Culture Jammers HQ by Justine Tunney and Micah White, two of the 'leaders' of the leaderless movement (Costanza-Chock, 2012). The piece called for a 'Tahrir moment' (Adbusters, 2011) and reflected on the type of direct action and strategy that characterized previous anti/alter-globalization social movements. In October 2011 the UK soon followed suit with its own similar call to action on the steps of St Pauls Cathedral in London. Alongside the original inception of Occupy Wall Street, the style and name of Occupy spread globally, with action taking place in over 1,500 cities (Occupy Wall Street, 2014). Attempts to contextualize and understand the significance of what might be learnt from the happenings that took place in New York, and the broader global Occupy movement until it's largely physical, but by no means imaginative, demise at the hands of state actors, has continued to steadily emerge from critical criminology and associated critical disciplinary strands of thought.

The past decade of newly energized resistance and contestation, of which only a handful of, arguably, Western-centric manifestations are listed here, has reignited discourse regarding how to meaningfully challenge an existing world order that is dominated by complex elite-driven ideologies. In many cases the necessary struggle for survival swallows whole the resistance or contestation agenda, but when that resistance and contestation manages to temporarily or partially extract itself from the neoliberal mire, the product is a complex, mystifying object. This object often only has small cracks in its opaque neoliberal residual coating that are willing or able to reveal the interior. For those seeking alternative futures, this time is fraught with tension and anxiety in ways that are reflected in past struggle, but also with added new complex dimensions that mystify as much as they inform.

In this chapter I combine a review of emerging literature on the Occupy movement with preliminary interviews that were conducted with Occupy members in the UK to explore state responses to Occupy. I employ a Gramscian view of the state and use the notions of coercion and consent, in their crassest and most commonly applied form, by arguing that methods of coercion (physical force) and consent (cultural persuasion) are employed by the state to stifle and silence contestation and dissent. Employing concepts that are derived from the work of Gramsci for analytical purposes is fraught with contested interpretation and competing discourses. However, insofar as this can be seen as problematic, it can also be seen as providing a flexible framework which can be engaged with in order to open up spaces for continued critical discussion. Within this contested rubric of Gramscian-based analysis and concepts, arguably one generalized thematic Gramscian concept can be upheld – that a successful contestation in some way 'depends on a satisfactory analysis of how this class itself holds power' (Ransome, 1992: 135). It is this commitment to analytically exploring the methods that are employed by the state to retain and sustain power that remains central to this chapter and holds the different elements of this work together. However, alongside a Gramscian-based commitment to identifying and unpicking contemporary manifestations of coercion and consent through an examination of the state response to Occupy, there is also a stronger connection between the themes. Essentially what weaves the elements of this chapter together is an endeavour and commitment to critical thinking about how the world works, and recognition of state power and how it arguably attempts to achieve this through silencing dissenting voices and through attempts to monopolize consciousness with its own agendas.

Recognition, privilege, diversity and finding commonality

Before engaging directly with the central themes and discussion of this chapter there are a few baseline considerations and acknowledgments to attend to. The positionality and case study subject matter of the researcher and the research are deeply embedded in a Western-centric, somewhat moderately privileged context. It is therefore the acknowledged limitation of this research that some of the content that is discussed makes generalizations that do not reach greater depths regarding the disproportionate and increased severity of silencing efforts by state actors, and likewise towards those with complex intersectional identities such as gender, sexuality and ethnicity, to name a few, who as a result can have vastly different experiences of state responses. The differences in response by state actors include, as an example, policing responses to protestors that put non-white activists at greater risk of jail time (Yassin, 2012: 95). To acknowledge this is of the utmost importance amid the efforts to identify common struggles. At the inception of Occupy Wall Street and the original construction of the working declaration about the occupation, the line 'as one people, formerly divided by the color of our skin, gender, sexual orientation, religion, or lack thereof, political party and cultural background...' was edited out as there was concern from camp members that the phrase would erase 'histories of oppression that marginalized communities have suffered' (Binh, 2011).

Furthermore, as much of the discussion here is supported with illustrative narratives from the Occupy movement and its affiliates, Occupy should be seen as one subset of resistance that is contained within a broader multitude of contemporary global anti-capitalist resistance movements. Its relationship is defined in solidarity with these global movement counterparts, but it is also a special case, with its immediate and long-term aims being deeply embedded in a specific, perhaps demonstrably Western, context that is in a distinct and 'radically different phase of anti-capitalist struggles' (Harvey, 2012: 119). As Fischer (2011) states, 'some have likened Occupy to the Arab Spring. That analogy suggests that Occupy will get the US military to turn on Washington and displace the federal government. Not too likely.' At the macrolevel these pluralities are visible not only through the distinction between global resistance movements that are visibly separate due to their noticeably different title that does not contain the word 'Occupy', but also within those that we understand to be a facsimile of the Occupy 'franchise'. Not all Occupy-affiliated or inspired actions are labelled 'Occupy', with the attached geographical location reference. Some of the related movements are sometimes referred to as the Occupy/Decolonize, Decolonize, Reclaim or (Un)Occupy movements (see Black Orchid Collective, 2012; Davis, 2011; Schrager Lang and Lang/Levitsky, 2012). This is in order to reflect the distinct struggles of Occupiers globally, for many of whom the term 'Occupy' is an extremely problematic word that is inevitably and inextricably tied to a history and continued legacy of colonialism. The term 'Decolonize' is sometimes chosen instead in recognition of the aforementioned point, and perhaps in recognition of contemporary manifestations of colonialism and the continued oppression of a number of populations and groups globally.

At the microsociological level the pluralities of resistance are also expressed in various lucid ways from within the Occupy movement. A narrative from Barksdale (2012) recounts the political makeup of the camp at Occupy Oakland describing a geography of disparate and distinct camps within the overall camp, with groups whose central advertised concerns covered a range of issues. The geographically observable plurality of resistance within the camps is also indicative of the pluralities contained within the wider 'left' resistance. In many respects we can apply a Foucauldian lens to these pluralities, and through the use of a Foucauldian interpretation of the relationship between power and resistance we can argue that 'for Foucault resistance occurs everywhere a power relation exists' (Worth, 2013: 39). If we understand the dispersal patterns of power to be complex – that is, that oppressive uses of power are not one-dimensional but acted out in multifarious forms – then we should understand that rather than a united singular homogenous resistance we are faced with 'a plurality of resistances, each of them a special case' (Foucault, 1998 [1979]:96 cited in Worth, 2013: 96) that is shaped by both global and local contexts (Uitermark and Nicholls, 2012).

Silencing dissent: Coercion, state violence and social harm

Ample evidence of coercive methods of physical force to control and halt the Occupy movement has well been documented in the emerging published literature: the use of tear-gas canisters and flash-bang grenades against protestors (Taylor, 2011: 138); the violent removal of the protestors from Zuccotti Park in New York (Writers for the 99%, 2011); and the now infamous 'Pepper Spray Cop', whose attack on seated student demonstrators 'quickly became the faces of liberal willingness to use violence against the Occupy/decolonize movement' (Schrager Lang and Lang/Levitsky, 2012: 225). There are many further available examples besides these illustrative few.

Further key themes emerging in terms of coercive state responses to the Occupy movement include, first, persistent surveillance of the Occupy camps by state actors. Barksdale (2012: 8) describes how the presence of CCTV did not go unnoticed by occupiers at Occupy LA and that there was a feeling of being perpetually watched. He states: 'there are cameras everywhere. You are being watched. You think you're occupying, but you're occupied.' Again this sentiment was echoed by occupiers at the Wall Street camp where it was noted that looming over the camp was 'the white cantilever of a mobile NYPD observation tower, maintaining a sinister Panopticon stare on the vista below' (Writers for the 99%, 2011: 63). Those involved with Occupy Boston also told how police distributed a series of leaflets that explained their intent to video-tape the occupation 'to better identify and prosecute' (Squibb, 2011: 174). The leaflets were distributed at the same time as police forces set about removing media entities from the area via claims that 'their safety could no longer be assured' (Squibb, 2011: 174). On the night of the eviction from Zuccotti Park, police at Occupy Wall Street arrested six journalists 'and barred countless others from entering to witness the scene' (Writers for the 99%, 2011: 179).

Second, a common form of state response to the Occupy movement has been regarding the use of public order rhetoric to criminalize the movement. Vitale (2011: 74), reflecting on policing at Occupy Wall Street, analysed the role that broken windows theory played and the impact that this had on policing in the USA whereby the police had become the actors who 'restore communities by controlling low level disorder' (Vitale, 2011: 74–75). He (2011) argues that the dissent of occupiers has been placed in the rubric of low-level disorder, as a break from the mundane and subsequently as a threat. Public order rhetoric has thus served as a stepping stone towards legitimizing the introduction of many protesters into the criminal justice system, leading to 'frequent arrests', such as those during the mobile Occupy protest that took to Brooklyn Bridge in October 2011 (Vitale, 2011: 80), and a total arrest count of over 830 people at Occupy Wall Street (Jaffe, 2011: 257). Within the literature it was also noted that the physical removal of many camps was based around some form of public-order rationale (Pickerill and Krinsky, 2012: 285). Gilmore (2010: 21) details how criminal justice institutions specifically make a 'critical distinction between "organized declared" and "non declared" protests'. This subsequently provides the conditions for a 'discourse of the dangerous, unpredictable, abnormal deviant[which] is a key foundation stone on which is built the culture of impunity and immunity surrounding state servants' (Sim, 2010: 6).

Within the emerging published narratives that might be considered as being captured under political society responses to the Occupy movement, it is of particular interest that clear distinctions between public and private policing and security were not always made. Separate from the Occupy movement, critical criminological-based discourses regarding the changes in trajectories of the nature of policing are better recognized to include 'private' security functions in collusion with public state forms of policing (see Button and John, 2002; Gillham et al., 2013; Morgan, 2014). Within the literature there was some recognition of the changing nature of the policing of protest in terms of the state–corporate relationship. For example, Pickerill and Krinsky (2012: 285) note that the 'politics of policing, especially in the collusion between financial interests and the repression of dissent, was made evident by the response to Occupy' and furthermore that 'Occupy has illustrated the extent to which protest policing has evolved'. However, the majority of the literature regarding Occupy makes ambiguous reference to 'the police', and the intricacies of its state–corporate manifestation are unclear. The recognition of composite 'grey policing' (Hoogenboom, 1991; Zedner, 2009) is recognized in only some of the literature (see Wolf, 2012; Bratich, 2014; Dellacioppa et al., 2013). For example, Bratich (2014: 68) makes reference to Bloomberg's "private army" who played a large role in the physical eviction of the Occupy Wall Street camp in New York, and Dellacioppa et al. (2013: 413) make reference to Occupy action on Skid Row in LA and reflect on how the area was 'also policed by private security, hired by the local business community'. In the former work, Bratich (2014: 68) analyses the complex state–corporate relationship further:

> Nationwide, public/private alliances were forged between local law enforcement, banks, private security firms, and federal agencies to spy on, restrict, and disrupt occupations. We saw the formation of 'fusion centers' where 'information sharing environments' were cultivated to enhance police powers.

Contrary to these limited discussions regarding the role of the corporate, it can be seen that the development and rationale for the Occupy movement include ample recognition and discussion regarding the role of the corporation in terms of sustaining economic inequality through the accumulation of wealth (see Linzey and Reifman, 2011; Declaration of the Occupation of New York City, 2011; Klein, 2011; Johnston, 2011; Kingsolver, 2011; Dixon, 2011; Phillips, 2012; Walia, 2011). However, despite some previously mentioned documented efforts to interrogate the state–corporate relationship, in comparison there appears to be limited critical discourse regarding the role of the corporate in the policing and silencing of the movement.

Interviews with Occupiers in the UK have reinforced concerns and the need for the continued problematizing of the role of actors from private corporations in the policing of contemporary protest movements. One interview participant recounted:

> It was a lot about private security guards as well. We'd done [occupied ... name of shop removed] not long before ... people were sitting on floor and security were kicking people in the ribs like they were really violent so people's confidence was really shaken ... I was punched in the face by a security guard and thrown outside. A pregnant woman was pushed to the floor by a security guard and then a few of the other protestors got in the way of the pregnant woman ... and just security got really heavy handed they got one lad in a headlock and dragged him off, put a cigarette out on his face, smashed his phone because he was filming what had taken place. Some of the Occupy activists phoned the police when the police turned up they just started arresting protestors.
>
> (Anonymous participant, Occupy Liverpool)

This is of increasing concern given already established private security trajectories and continued speculation *vis-à-vis* the increased privatization of policing (Taylor and Travis, 2012), the implications of which might include a lack of access to data under the Freedom of Information Act 2000 (Gilmore, 2010), concerns about corporate accountability and the ability of corporations to enhance, rather than replace, blocs of state power (Tombs and Whyte, 2009).

Silencing dissent: Consent and pacification

Žižek (2012: 83) described the Wall Street protests as a 'formal gesture of rejection', a sentiment in which they are not alone. Fischer (2011) describes the scene from the 1976 film *Network* where 'a televisions news anchorman yells out on the air, "I'm as mad as hell, and I'm not going to take this anymore!" He gets thousands, maybe millions, of Americans to open their windows and yell the same phrase into the streets below. Felt good, changes nothing.' The depiction of this scene is only one-dimensional; to what extent does it adequately illustrate Occupy in its physical presence (on camp) and virtual presence (off camp and online)?

In an age of changing and diverse media for interactions and engagement with social movements, there are questions to be asked regarding the role that technologies play in the format and *modus operandi* of protest and dissent.

Lim (2013: 637) highlights the recent employment of terms such as 'Slacktivism', 'Clicktivism' and 'armchair activism' which are used to express cynical views of engagement with activism from behind or through the computer screen or mobile device, where participation takes place in the virtual world and is not always translated into action in physical spaces. What is noticeable is the number of 'likes' on Facebook – and for that matter followers on Twitter, although Facebook appears to have been the most popular social media format for Occupy (Gaby and Caren, 2012: 369) – compared with the number of people taking part in the Occupy camps. Similar disparities can be seen for most Occupy camps and Facebook groups globally. For example, in the Liverpool UK camp at the height of engagement there were approximately 40 people involved in the physical Occupy campsite within the urban centre, compared with over 5,000 'likes' of the Occupy Liverpool group on Facebook.

While there is arguably some truth in claims that 'social media activisms are always in danger of being too fast, too thin and too many' (Lim, 2013: 654), it is important to note that social media has 'been critical for linking potential supporters in order to share information and stories' (Gaby and Caren, 2012: 367) within and between local Occupy sites and networks of Occupiers globally. The Occupy movement also emanated from an online call to action and has been endorsed by Anonymous hacktivists (Costanza-Chock, 2012), who have long since drawn attention to the ability to dissent and disrupt in the virtual world and not just the physical world. Thus, rather than seeing engagement with online activities related to activism as 'subordinate to "real" (physical) activism' (Lim, 2013: 637), it can be seen as complimentary supporting various activities such as organization, information dissemination and coordination for offline activities (Ayres, 1999; Bennett et al., 2008, cited in Christensen, 2011: 2), with some even arguing so far as to say that 'the Internet has a positive impact on off-line mobilization' (Christensen, 2011: 1). Such usage contests and reclaims technological devices in the face of a market that has arguably 'allowed capitalism to conflate desire and technology so that the desire for an iPhone can now appear automatically to mean a desire for capitalism' (Fisher, 2012: 131).

However, in terms of assessing the potential trajectory of technology and social media usage for activist-related activities, there are established concerns regarding the importance of actually taking to the streets into physical space, including the 'counter-temporality' (Adams, 2011, cited in Halvorsen, 2012: 431) that is achieved through disrupting and reclaiming spaces of capitalist production, such as those that have been seen in mobile occupations in and around major cities. In fact, state responses to activist activities, both online and offline, offer some thought-provoking revelations, such as the response of the four-day Internet 'blackout' in Egypt during the events of 2011 in Tahrir Square, when online practices were translated into action on the streets. Despite repeated revelations regarding intensified online surveillance practices from the state regarding the online activities, objectionable water cannon purchases remain reserved for those dissenting in public, not those at their computer screens. Brown (2013) explains how online activities 'allow people to participate from a distance, to dip a toe into what is, for those on the ground, a dangerous environment' which, given the relentless state authoritarian response to protestors, can only be to the convenience of the state, which is able to push back, pacify and ultimately jostle to consign dissent into intangible 'likes' and 'tweets' on personal computer devices.

Coercion and consent: Criminalizing dissenters and the curious role of 'the other'

Although discussed previously in isolation, coercion and consent are in reality a dialectical strategy (Ransome, 1992: 135), working in 'synthesis' (Thomas, 2012: 49). An example that is worthy of attention emerged during an interview in my primary research with a participant from Occupy London, where the discussion turned to drug and alcohol use on camp at St Pauls. During the protest there was concern regarding the use of drugs and alcohol that manifested itself in the Occupy Safer Spaces Policy where, alongside other banned behaviour, such as 'racism, ageism, homophobia, sexism, transphobia, ableism or prejudice based on ethnicity, nationality, class, gender, gender presentation, language ability, asylum status or religious affiliation', it was declared that '13. Occupy London is an alcohol and drugs free space.' The Safer Spaces Policy listed 13 points that were aimed at providing guidance so that Occupy London could 'operate and conduct [our] discussions in a safe anti-oppressive space...that is welcoming, engaging and supportive' (OccupyLondon.org,uk). According to Occupy London.org.uk, it was felt that in order to ensure that the aforementioned 'safe space' was created, it was 'necessary to establish some guidelines for participants'. At a basic level, arguably one can see that drug and alcohol consumption on camp would need to be discussed, not least in terms of safeguarding participants in the Occupy movement whose consumption of drug and alcohol might be prosecuted by state actors. However, concerns about the responses by the movement towards those with drug and alcohol (mis)use issues were expressed:

> But I remember that one of the key moments was ... I mean basically the unleadership of Occupy wanted to pass this so called safer spaces policy, I mean it was a big issue and a big thing ... it was all about how we were going to maintain good behaviour and then the last tiny little thing said 'Occupy London is a drug and alcohol free space' and I thought is it? And how can you make a space full of adults alcohol free in a country that legalises it anyway it doesn't make sense right ... I was basically saying look first of all that is clearly not true, secondly it's a great intention but it is unrealistic and it will alienate some people so wouldn't it be better if we say ... we would like it to be this way but if you need help you know we are ...
>
> (Anonymous participant, Occupy London)

If Occupy were to admit that there may have been drug and alcohol users on camp, it would have run the risk of a mainstream media attack, much of which was already keen to demonize and 'discredit' the movement. However, when reflecting on the potential exclusion or denial of persons through the safer spaces policy list of banned actions that included no drink or drug use on camp, it raised a poignant question: To what extent does, or did, drug use have on people and their access to the imaginary yet tangible 'legitimate occupy protestor' status? Successful deployment of consent by the ruling elite can be recognized to a greater extent in the external responses to the protest movement. For example, easily identifiable manifestations of the successful deployment of consent can be seen in the 'get a job' cat calls from the passers-by who were external to the movement. It is now we turn our attention to the more subtle and nuanced manifestations of a partially successful deployment of consent 'within' the resistance. To extend the Gramscian notion of 'active hegemony' (Thomas, 2012), elements of the resistance can themselves become active hegemonists, reproducing popular layers of consent. Thomas (2012) argues that there is an energizing dimension to such active hegemonic-based tactics. To what extent are those whose intentions are to contest actually and effectually co-opted into the frame to do the bidding of the bourgeoisie? The critique of the internal responses to people within the camps is not a critique of the Occupy movement itself but instead a critical exploration of the deployment of the civil society arm of the state through the exploitation of strategies of consent within the movement.

Something else that is worthy of attention is a consideration of the impact of labels such as 'homeless', 'drug user' or 'alcohol user', and the depoliticizing effect that this can have. A good example of this is regarding the occupation at Finsbury Square camp in London where *Guardian* writer Paul Walker described the makeup of the camp, stating: 'the longer it went on it attracted an increasing number of vulnerable homeless people, often with drink or drug problems, *rather* than protesters' (Walker, 2012, author's emphasis). The popular conception here suggests that one cannot be someone who has 'drink and drug problems' and also be a protestor. Ergo the ability to engage politically is reducible to, and restricted to, a litmus test of political agency. This test serves only to reinforce discriminatory ideologies regarding the 'status' and abilities of those with alcohol and drug (mis)use issues to be politically active. In this case the label of 'drug user' becomes a 'master status' (Becker, 1963), and as such it depoliticizes and fails to recognize that one can be a drug user and a protestor. Thus there is a construction and expansion of the legitimate and non-legitimate protestor which is reproduced even by more critical sources. Of course, the Invisible Committee (2009: 37) states that 'the other' holds a curious position. Using the illustration of immigrants in France, it states that 'immigrants assume a curious position of sovereignty: if they weren't here, the French might stop existing'.

Masquelier (2013: 404) describes how Occupy is arguably 'wholly distrustful of the rules of conduct with which society as presently constituted provides each of its members'. However despite this claim, through the language and actions undertaken via the safer spaces and zero tolerance policies, questions have been raised regarding the adoption of procedures that are associated with the state. The language of safer spaces and zero tolerance derive from Rudy Giuliani's policing strategy legacies from the 1990s and similarly New Labour-esque strategies that are influenced by the aforementioned US approach to policing. Grande (2013: 369) argues that the occupation of space can reinforce 'settler hegemony'. In this case it can be applied to the Occupy settlers and their claims to that space, and the application of new rules that then displaces groups such as the homeless from spaces that they have long since occupied. Similar sentiments can be found in other sites beyond Occupy Wall Street where one Occupier called the displacement of the homeless in El Paso as a form of 'colonization'. Smith et al. (2012: 360) note:

> If we insist on controlling the homeless people's actions and impose our rules on them then we become colonizers...we become a greedy corporation.
>
> (Occupy El Paso Facebook post cited in Smith et al., 2012: 360)

Further questions are raised in this area by Chadeyane Appel (2011: 120), who discusses how some homeless individuals, in order to use the space in McDonalds, were required to make a US$ 2 contribution for every 12 hours spent in the building in order to be present, and that Occupy in some respects is similarly reproducing this strategy by ensuring that one must 'contribute' to the movement in order to be allowed access. Such analogies are evocative of strong tendencies towards the reduction in 'free' space, and they arguably raise questions regarding to what extent elements of Occupy reproduced spaces of consumption and exclusionary practices in the neoliberal city (Coleman and Sim, 2010; Davis, 1992; Lefebvre, 1996). The following quote from an interview with a member of Occupy Liverpool reflects these concerns further:

> Like a couple of homeless guys turned up the morning after the first night and they just always come and sit on the monument and have a butty [sandwich] from the hostel and a can of beer and it was sort of like not so much explaining to them that you can't come and drink beer here, because we're going this is a no drinking camp, it was more the issue was explaining to other people on the camp that those guys do that every day. They've done that since before we were there, like who are we to tell them they can't.
>
> (Anonymous participant, Occupy Liverpool)

Schein (2012: 339) states, 'Occupiers variously resisted and succumbed to a language dividing the "real" political occupiers from those drawn to the park by the 'promise of a real meal and a safe space to sleep'. The aforementioned decisions appeared to be made on the basis of an 'informal contribution calculus' (Herring and Glück, 2011: 166) regarding (a problematically and subjectively defined) 'contribution' to the camp which determined if you were welcome there or deemed to be present for 'other' purposes. Herring and Glück (2011) regard this informal calculus as being an internal manifestation of the bourgeois concept of the deserving and undeserving poor. The examples above seek to better explain the conditions in which the Occupy movement and similar contemporary forms of contestation operate. This is not to say that this is always the case because some Occupy sites, such as Occupy

LA, focused much of its activist work on 'LA's Skid Row residents and organizations to fight against gentrification and the criminalization of poverty' (Dellacioppa et al., 2013: 304). However, social movements are complex entities and previous analytical failures have been highlighted by Piven and Cloward (1978: x), where there has been a failure 'to understand that the main features of contemporary popular struggles are both a reflection of an institutionally determined logic and a challenge to that logic'. The title of this paper employs the term 'capitalist realism', which was employed by Fisher (2009) to in some way explain the current condition of what Fisher finds to be a distinct lack of alternatives to capitalism. He (2009: 7) argues that we are dealing with 'a deeper, far more pervasive, sense of exhaustion, of cultural and political sterility'. Part of this political sterility and lack of alternative stems from the all-pervasive cultural hegemony that is arguably maintained by the deployment of coercion and consent that seeps into all of our endeavours, including attempts to dissent from and contest the *status quo*.

Conclusion: Transformations in consciousness, self-realization and persistence

A meaningful challenge to the capitalist *status quo* is arguably one that challenges the self-styled related ideologies that are imposed through methods of state-deployed coercion and consent. A quick and easy solution to this is not forthcoming, nor is it offered or to be found within the remit of this chapter. Instead I suggest that greater attention and critical discussion might be warranted to unveil the hegemonic methods of the powerful, not only outside the resistance but within it. Harvey (2012: 122) demonstrates that we can see examples of resistance and contestation that in fact mimic what they seek to contest, such as cooperative enterprises that 'tend at some point to mimic their capitalist competitors'. This is not to say that this is not attempted by others, but it is through fear of coercive reprisal that meaningful analyses of forms of consent 'within the ranks' are less often challenged. Arrighi et al. (1989: 29) state that 'opposition is permanent, but for the most part latent. The oppressed are too weak – politically, economically and ideologically – to manifest their opposition constantly.' This chapter aims to provide an illustrative insight into contemporary hegemony in action through coercion and consent that are levelled at the resistance from outside actors but also present within the resistance. Despite the difficulties in maintaining opposition constantly, part of that contestation is to hold a mirror up to and reflect on a number of practices within all of our resistance efforts that are often dripping with the ideological coatings that they seek to contest.

Cader (2013) states that 'our horizons of change seem limited', especially when we consider that meaningful anti-capitalist politics require breaking free from a series of limitations and restrictions, and subsequently seeking ways to 're-potentialise' the world (The Free Association, 2011). Brissette (2011) states that 'we must deny the existing system the power to define the situation for us', but this is easier said than done because 'a particular problem is the duality of the oppressed: they are contradictory, divided beings, shaped by and existing in a concrete situation of oppression and violence' (Freire, 2996: 37). Nader (2011: 75), describing scenes from Occupy, suggests that 'the campers and the marchers are discovering that they have power – the crucial first stage of liberation from growing up powerless and under corporate domination – the two go together – into a process of self-realization'. Actualization of self-realization is no easy task when coercion and instilling fear in terms of reprisal of coercive methods continue to be an omnipotent and omnipresent feature of the activist life.

As hooks (1994: 78) states, 'progressive politics must include a space for rigorous critique, for dissent, or we are doomed to reproduce in progressive communities the very forms of domination we seek to oppose'. Such spaces for rigorous critique are under constant assault from established and newer-evolving manifestations of coercion and consent from a persistently 're-codifying' and 'ambidextrous' state (Coleman et al., 2009; Peck, 2010). Social movements, activism, protest and resistance remain key areas to turn our attention to, especially in terms of exploring the responses to them and those within them as a means to critique hierarchies of power, foster emancipatory knowledge and challenge existing power relations.

Bibliography

Adbusters (2011) Occupywallstreet: A shift in revolutionary tactics. Accessed on 24 June 2014 at https://www.adbusters.org/blogs/adbusters-blog/occupywall street.html.

Arrighi, G., Hopkins, T.K. and Wallerstein, I. (1989) *Anti-Systemic Movements*. London: Verso.

Barksdale, A. (2012) Occupy LA: The worst and best journal of communist theory and practice 5. Accessed on 26 February 2012 at http://insurgentnotes .com/.

Becker, H.S. (1963) *Outsiders: Studies in the Sociology of Deviance*. New York: The Free Press.

Binh, P. (2011) The nuts and bolts of #OccupyWallStreet. Accessed on 29 May 2014 at http://www.indypendent.org/2011/09/29/nuts-and-bolts-occupywallstreet.

Black Orchid Collective (2012) The radicalization of decolonize/occupy Seattle (Guest Article), *Journal of Communist Theory and Practice*, 5. Accessed on 26 February at http://insurgentnotes.com/.

Bratich, J. (2014) Occupy all the dispositifs: Memes, media ecologies, and emergent bodies politic, *Communication and Critical/Cultural Studies*, 11(1): 64–73.

Brissette, E. (2011) For the fracture of good order. Accessed on 16 April 2012 at http://bjsonline.org.

Brown, J. (2013) From activism to occupation. Accessed at http://currents.cwrl.utexas.edu/2013/from-activism-to-occupation.

Button, M. and John, T. (2002) 'Plural Policing' in action: A review of the policing of environmental protests in England and Wales, *Policing and Society: An International Journal of Research and Policy*, 12(2–1): 111–121.

Cader, A. (2013) Towards Revolutionary Unity. Accessed on 15 March 2013 at http://anticapitalists.org.

Castañeda, E. (2012) The Indignados of Spain: A precedent to Occupy Wall Street, social movement studies, *Journal of Social, Cultural and Political Protest*, 11(3–4): 309–319.

Chadeyane Appel, H.C. (2011) 'The bureaucracies of anarchy'. In Schrager Lang, A. and Lang/Levitsky, D. (eds) *Dreaming in Public: Building the Occupy Movement*. Oxford: New Internationalist Publications, pp. 112–120.

Christensen, H.S. (2011) Political activities on the Internet: Slacktivism or political participation by other means?, *First Monday*, 16(2). Accessed at firstmonday.org.

Coleman, R. and Sim, J. (2010) Managing the mendicant: Regeneration and repression in Liverpool, *Criminal Justice Matters*, 92(1): 30–31.

Coleman, R., Sim, J., Tombs, S. and Whyte, D. (2009) *Introduction: State Power Crime*. London: Sage, pp. 1–19.

Costanza-Chock, S. (2012) 'Mic check! Media cultures and the Occupy movement', *Social Movement Studies: Journal of Social, Cultural and Political Protest*, 11(3–4): 375–385.

Davis, M. (1992) 'Fortress Los Angeles: The militarization of urban space', *Variations on a Theme Park*, pp. 154–180.

Davis, A. (2011) *(Un)Occupy*. In Taylor, A., Gessen, K. and editors from n+1, Dissent, Triple Canopy and The New Enquiry (eds) *Scenes from Occupied America*. London: Verso, pp. 132–133.

Dellacioppa, K., Soto, S. and Meyer, A. (2013) 'Rethinking resistance and the cultural politics of occupy', *New Political Science*, 35(3): 403–416.

Dixon, B.A. (2011) 'Occupy where? What's in it for black and brown people'. In Schrager Lang, A. and Lang/Levitsky, D. (eds) *Dreaming in Public: Building the Occupy Movement*. Oxford: New Internationalist Publications, pp. 143–146.

Fischer, C. (2011) Occupy! Now what? Made in America: Notes on American life from American History. Accessed on 16 April 2012 at http://bjsonline.org.

Fisher, M. (2009) *Capitalist Realism: Is There No Alternative?* London: Zero Books.

Fisher, M. (2012) 'Chapter 13: Post-capitalist desire'. In Campagna, F. and Campiglio, E. (eds) *What Are We Fighting For: A Radical Collective Manifesto*. London: Pluto Press, pp. 131–138.

Freire, P. (1996) *Pedagogy of the Oppressed*, 2nd edn. London: Penguin.

Gaby, S. and Caren, N. (2012) 'Occupy online: How cute old men and malcolm X recruited 400,000 US users to OWS on Facebook', *Social Movement Studies*, 11(3–4): 367–374.

Gillham, P.F., Edwards, B. and Noakes, J.A. (2013) 'Strategic incapacitation and the policing of Occupy Wall Street protests in New York City, 2011', *Policing and Society: An International Journal of Research and Policy*, 23(1): 81–102.

Gilmore, J. (2010) 'Policing protest: An authoritarian consensus', *Criminal Justice Matters*, 82: 21–23.

Grande, S. (2013) 'Accumulation of the primitive: The limits of liberalism and the politics of Occupy Wall Street', *Settler Colonial Studies*, 3(3–4): 369–380.

Halvorsen, S. (2012) 'Beyond the network? Occupy London and the global movement, social movement studies' *Journal of Social, Cultural and Political Protest*, 11(3–4): 427–433.

Harvey, D. (2012) *Rebel Cities: From the Right to the City to the Urban Revolution*. London: Verso.

Herring, C. and Glück, Z. (2011) 'The homeless question'. In Taylor, A., Gessen, K. and editors from n+1, Dissent, Triple Canopy and The New Enquiry (eds) *Scenes from Occupied America*. London: Verso, pp. 163–169.

Hoogenboom, B. (1991) 'Grey policing: A theoretical framework', *Policing and Society: An International Journal of Research and Policy*, 2(1): 17–30.

hooks, b. (1994) *Outlaw Culture: Resisting Representations*. Oxon: Routledge.

Jaffe S. (2011) 'Occupy Wall Street prepares for crackdown: Will Bloomberg try to tear it down?'. In Schrager Lang, A. and Lang/Levitsky, D. (eds) *Dreaming in Public: Building the Occupy Movement*. Oxford: New Internationalist Publications, pp. 254–259.

Johnston, A. (2011) 'What I saw at Occupy Wall Street last night, and what I saw when I left'. In Schrager Lang, A. and Lang/Levitsky, D. (eds) *Dreaming in Public: Building the Occupy Movement*. Oxford: New Internationalist Publications, pp. 75–78.

Kingsolver, B. (2011) 'Another American way'. In Schrager Lang, A. and Lang/Levitsky, D. (eds) *Dreaming in Public: Building the Occupy Movement*. Oxford: New Internationalist Publications, pp. 73–74.

Klein, N. (2011) 'The most important thing in the world'. In Schrager Lang, A. and Lang/Levitsky, D. (eds) *Dreaming in Public: Building the Occupy Movement*. Oxford: New Internationalist Publications, pp. 43–46.

Lefebvre, H. (1996) *Writings on Cities*. Oxford: Blackwell.

Lim, M. (2013) 'Many clicks but little sticks: Social media activism in Indonesia', *Journal of Contemporary Asia*, 43(4): 636–657.

Linzey, T. and Reifman, J. (2011) 'Chapter 13: How to put the rights of people and nature over corporate rights'. In van Gelder, A. and the staff of YES! Magazine (eds) *This Changes Everything: Occupy Wall Street and the 99% Movement*. San Francisco: Berrett-Koehler Publishers, pp. 70–73.

Masquelier, C. (2013) 'Critical theory and contemporary social movements: Conceptualizing resistance in the neoliberal age', *European Journal of Social Theory*, 16(4): 395–412.

Morgan, M. (2014) 'The containment of occupy: Militarized police forces and social control in America', *Global Discourse: An Interdisciplinary Journal of Current Affairs and Applied Contemporary Thought*, 4(2–3): 267–284.

Nader, R. (2011) 'Chapter 14: Going to the streets to get things done'. In Van Gelder, S. (ed.) *This Changes Everything: Occupy Wall Street and the 99% Movement*. San Francisco: Berrett-Koehler Publishers (Yes! Magazine), pp. 74–76.

Occupy London Safer Spaces Policy. Accessed on 20 July 2014 at http://occupylondon.org.uk/about/statements/safer-space-policy/.

Occupy Wall Street. Accessed on 20 July 2014 at http://occupywallst.org/.

Occupy Wall Street (2011) 'Declaration of the occupation of New York City'. In Schrager Lang, A. and Lang/Levitsky, D. (eds) *Dreaming in Public: Building the Occupy Movement*, Oxford: New Internationalist Publications, pp. 49–51.

Peck, J. (2010) 'Zombie neoliberalism and the ambidextrous state', *Theoretical Criminology*, 14: 104–110.

Phillips, M. (2012) 'Room for the poor'. In Schrager Lang, A. and Lang/Levitsky, D. (eds) *Dreaming in Public: Building the Occupy Movement*. Oxford: New Internationalist Publications, pp. 270–276.

Pickerill, J. and Krinsky, J. (2012) 'Why does Occupy matter?' *Social Movement Studies: Journal of Social, Cultural and Political Protest*, 11(3–4): pp. 279–287.

Piven, F.F. and Cloward, R.A. (1978) *Poor People's Movements: Why They Succeed, Why They Fail*. New York: Random House.

Ransome, P. (1992) *Antonio Gramsci: A New Introduction*. Hertfordshire: Harvester Wheatsheaf.

Schein, R. (2012) 'Whose occupation? Homelessness and the politics of park encampments', *Social Movement Studies: Journal of Social, Cultural and Political Protest*, 11(3–4): 335–341.

Schrager Lang, A. and Lang/Levitsky, D. (2012) 'Introduction: The politics of the impossible'. In Schrager Lang, A. and Lang/Levitsky, D. (eds) *Dreaming in Public: Building the Occupy Movement*. Oxford: New Internationalist Publications, pp. 15–25.

Sim, J. (2010) 'Thinking about state violence', *Criminal Justice Matters*, 82(1): 6–7.

Smith, C., Casteñada, E. and Heyman, J. (2012) 'The homeless and Occupy El Paso: Creating community among the 99%', *Social Movement Studies: Journal of Social, Cultural and Political Protest*, 11(3–4): 356–366.

Squibb, S. (2011) 'Scenes from occupied Boston'. In Taylor, A., Gessen, K. and editors from n+1, Dissent, Triple Canopy and the New Inquiry (eds) *Scenes from Occupied America*. London: Verso, pp. 170–175.

Taylor, M. and Travis, A. (2012) G4S chief predicts mass police privatisation. Accessed on 20 June 2014 at http://www.theguardian.com/uk/2012/jun/20/g4s-chief-mass-police-privatisation.

Taylor, S. (2011) Scenes from Occupied Oakland'. In Taylor, A., Gessen, K. and editors from n+1, Dissent, Triple Canopy and the New Inquiry (eds) *Scenes from Occupied America*. London: Verso, pp. 134–145.

The Free Association (2011) 'On fairy dust and rupture'. In Lunghi, A. and Wheeler, S. (eds) *Occupy Everything: Reflections on Why It's Kicking Off Everywhere*. Brooklyn: Minor Compositions, pp. 24–31.

The Invisible Committee (2009) *The Coming Insurrection*. London: MIT Press.

Thomas, M. (ed.) (2012) *Antonio Gramsci Working Class Revolutionary: Essays and Interviews*. London: Workers Liberty.

Tombs. S. and Whyte, D. (2009) 'The state and corporate crime'. In Coleman, R., Sim, J., Tombs, S. and Whyte, D. (eds) *State Power Crime*. London: Sage, pp. 103–115.

Uitermark, J. and Nicholls, W. (2012) How local networks shape a global movement: Comparing Occupy in Amsterdam and Los Angeles, *Social Movement Studies: Journal of Social, Cultural and Political Protest,* 11(3–4): 295–301.

Vitale, A. (2011) 'NYPD and OWS: A clash of styles'. In Taylor, A., Gessen, K. and editors from n+1, Dissent, Triple Canopy and the New Inquiry (eds) *Scenes from Occupied America.* London: Verso, pp. 74–81.

Walker, P. (2012) Occupy Finsbury Square Camp Removed. Accessed on 29 May 2014 at http://www.theguardian.com/uk/2012/jun/14/occupy-finsbury-square-camp-removed.

Walia, H. (2011) 'Letter to Occupy Together movement'. In Schrager Lang, A. and Lang/Levitsky, D. (eds) *Dreaming in Public: Building the Occupy Movement,* Oxford: New Internationalist Publications, pp. 164–170.

Wolf, N. (2012) Revealed: How the FBI coordinated the crackdown on Occupy: New documents prove what was once dismissed as paranoid fantasy: totally integrated corporate-state repression of dissent, *Guardian.* Accessed on 12 August 2013 at http://www.theguardian.com/commentisfree/2012/dec/29/fbi-coordinated-crackdown-occupy.

Worth, O. (2013) *Resistance in the age of austerity: Nationalism, the Failure of the Left and the Return of God,* New York: Zed Books.

Writers for the 99% (2011) *Occupying Wall Street: The inside Story of an Action that Changed America.* London: OR Books.

Yassin, J.O. (2012) 'Occupy Oakland day four: Wherein I speak to some folks and the general assembly debate move on's move in'. In Schrager Lang, A. and Lang/Levitsky, D. (eds) *Dreaming in Public: Building the Occupy Movement.* Oxford: New Internationalist Publications, pp. 92–98.

Zedner, L. (2009) *Security.* Oxon: Routledge.

Žižek, S. (2012) 'Chapter 7: Occupy Wall Street, or, the violent silence of a new beginning'. In Žižek, S. (ed.) *The Year of Dreaming Dangerously.* London: Verso, pp. 77–91.

13
Refugee Protests and Political Agency: Framing Dissensus through Precarity

Katrin Kremmel and Brunilda Pali

Introduction

Increasingly, refugees, asylumseekers and *sans papiers* have in recent decades engaged in a large number of protests throughout Europe (Pojmann, 2008). The occupation of the Parisian church Saint-Bernard by *sans papiers* in 1996, the foundation of the 'Caravan of Migrants' in Germany in 1997, the month of protests that were organized in Northern Italy in the spring of 2001, the national demonstrations of the *sans papiers* in Switzerland in 2001, and the refugee protests and occupations taking place since 2012 in Austria are all examples of political actions, sometimes creating common political platforms. Since 2012, refugee protests have also occurred in Denmark, Turkey, Bulgaria, Greece and the Netherlands. A Refugee Congress was held in March 2013 in Munich, Germany, where participants discussed future actions, organizational bodies and a list of political demands aiming to change the conditions for refugees and asylumseekers in Europe.

Predominantly, these protests have been framed by media and politicians as disturbances of the national public order, while their subjects – refugees, asylumseekers, *sans papiers* – and their supporters have been constructed as 'dangerous' agents, cast as the object of fears and anxieties that are produced by discursive processes of securitization and criminalization. These depictions strongly contrast and contradict the recurring discursive image of 'the refugee' as a humanitarian subject in need, waiting to be rescued, cared for and protected. Among various coexisting discourses on immigrants, refugees and asylumseekers, there is certainly a recognizable ambivalence between two prevalent discourses which often appear side by side: on the one hand, the humanitarian discourses that portray them as human beings in need of being cared for (Bousfield, 2005; Anderson, 1996; Fassin, 2007, 2012; Papastergiadis, 2006; Lavenex, 2001; Rajaram, 2002; Zetter, 1991), and on the other hand, the security discourses portraying them as a 'problem' or 'deviant' population that constitutes a threat to national belonging, security and order (Buonfino, 2004; Zierer, 1998; Pickering, 2001; Lammers, 1999; Nyers, 2003; Castañeda, 2010; Shane Boyle, 2011; Aas, 2011; Aas and Bosworth, 2013). Framing the refugee as a human-itarian object certainly has different implications from framing them as an object of security, but what unites both of these discourses is the depoliticization of the refugee as a subject, who is presented as speechless. The refugee is silenced through representation, as attempts of self-representation are hardly being acknowledged as such by various social actors. In other words, the paradox of the 'refugee discourses' is that they have no place for its very subject (Soguk, 1999).

We oppose these depictions by emphasizing a reading of the refugee movements, which has already been forwarded by the protesting refugees themselves, as political acts of self-representation (see Cissé, 1999; Refugee Protest Camp Vienna, 2014; The VOICE Refugee Forum Germany, 2014; WIJZIJNHIER, 2014). We will thus read the refugee movements, focusing on the case of the Austrian refugee protests, which started in 2012, by framing the refugees as political agents, and their actions as constituting political action. We will use the concept of pol-itics as dissensus, as defined by Jacque Rancière, as a way to frame political subjectivity as the action of taking agency instead of having agency. We also make use of the concept of precarity, as used by Judith Butler, as a basis for a social ontology of political action. We argue that this concept offers a way out of notions of victimhood, while retaining the central focus on the precarity and vulnerability of the refugees.

We will use Hannah Arendt's work on statelessness as a starting point for this chapter and its theoretical discussion. It is almost impossible to speak of the refugee without thinking of her contribution (Benhabib, 2004; Borren, 2008; Bousfield, 2005; Cotter, 2005; Krause, 2008; Oman, 2010). We have found especially useful Krause's (2008) rereading of Arendt in the context of the *sans papiers* – undocumented economic migrants in France who have engaged in highly visible protests. Krause applies Arendt's themes of rightlessness but also shows how the *sans papiers* have been able to subvert their situation and call upon a wider public to support their cause. Following Arendt, Krause argues that undocumented migrants are victims of total domination. However, she suggests that they also have the potential to be political actors 'whose public appearance can be explosive and liberating' precisely because '[he] embodies the contradictions of the arrangements that exclude him' (Krause, 2008: 341).

While we do agree with the substance of this assessment and we find it very useful at an empirical level, we came to the conclusion that Arendt's work does not provide the foundation to support it on a theoretical level, given that in her thinking it is simply a conceptual impossibility for the rightless to claim the 'right to have rights'. In the following section we will make an effort to explain our apprehensions by focusing on substantial points of Arendt's understanding of political action.

The 'Refugee protest camp Vienna' movement

On 24 November 2012, a group of more than 100 asylumseekers and supporters departed on a 35 km protest march from the reception centre for refugees in Traiskirchen to Vienna.[1] The march, followed by the refugee protest camp which was set up in Sigmund-Freud Park where the demonstrating refugees settled down after their arrival in the Austrian capital, was the starting point of a political engagement of refugees that was without precedent in Austria. From 24 November 2012 onwards the camp turned into a freely accessible and open space for political discussions and activities, mostly carried out in working groups and during daily plenaries in the evenings. It also turned into a 'meeting point', where people sat down together and talked over a cup of coffee and self-made food. Compared with the living conditions in Traiskirchen, the camp allowed for a relatively high degree of self-determination, since the refugees were not subjected to the forms of constant control and institutional restrictions which they had experienced on a daily basis at the reception centre.

On 18 December, after weeks in the refugee protest camp, and after weeks of being ignored by the government, the group decided to move into the church that bordered the protest camp. That morning the refugees formed little groups, which entered the church one after the other, sat down on the benches and waited quietly until everybody had arrived. Hoping to regain visibility and to increase the pressure on the government to respond to their demands, they were about to announce the first church occupation in Austrian history, when they decided on a last-minute call to term it 'seeking shelter' instead, aware of the risk that they were running otherwise to provoke fantasies that were so often referred to by the FPÖ[2] of migrants invading Austria, occupying national territory, and competing for the resources of the welfare state and on the labour market.

When the asylumseekers sought refuge in the church, the government could no longer ignore them. On 21 December, a group of refugees was received by a committee consisting of members of the government but only to be told that nothing could be done for them.

Nevertheless, moving into the church brought about a few changes that nobody had expected. The church authorities, which were quite ambivalent about their new role as hosts for about 50 refugees, appointed the Caritas to take care of the people inside the church and hired a private security service to control the church's entrances. The public were thus prohibited from entering freely, and the freedom of movement of the refugees was limited, as they and five supporters were allowed to move in and out only until the gates were closed at 10:00 pm. After these restrictions had been introduced, many refugees remarked that the situation inside the church had begun to resemble what life had been like in Traiskirchen.

At 4:00 am on 28 December, a large number of police violently evicted the camp. They claimed that an administrative law of the city, the Kampierverbot (prohibition of camping), had been violated. Both the legal grounds and the proportionality of the undertaken measures were heavily criticized as an illegitimate and disproportionate demonstration of state power. Frustrated with these developments, the refugees decided to go on hunger strike and successfully forced the government to continue negotiations. Although it did not respond positively to any of the refugees' demands, the movement had achieved what other civil society actors had not managed to do during recent decades: to turn the attention of the whole nation onto Austrian asylum policies.

From no 'right to have rights' to political action?

In *The Origins of Totalitarianism*, Arendt argues that the universal rights of man by virtue of membership of the human race, granted either in the nature of man or in God, were unattainable and of little value without citizenships rights that are guaranteed by a body of a nation state or political community that is able to enact and protect them. She argues through the example of people who were deprived of their citizenship that, without political and national rights, 'the stateless' are reduced to the 'abstract nakedness of being human' (Arendt, 1958: 299). 'If a human being loses his political status', Arendt writes, 'he should, according to the implications of the inborn and inalienable rights of man, come under the situation for which the declarations of such general rights provided.' But what the experiences of the refugees of 1943 revealed to her is precisely the opposite. It seemed to her 'that a man who is nothing but a man has lost the very qualities which make it possible for other people to treat him as a fellow-man' (Arendt, 1958: 300).

Thus follows the apprehension that

> We are not born equal; we become equal as members of a group on the strength of our decision to guarantee ourselves mutually equal rights. Our political life rests on the assumption that we can produce equality through organization, because man can build, act in and change a common world, together with his equals and only with his equals.
>
> (Arendt, 1958: 301)

Arendt's insight is that the crisis of the 20th century has taught us the fallacy and naiveté of believing that human rights can be defended by legal means alone; putting human rights into practice depends on something other than mere legal or formal structures. If human rights are possible at all for Arendt, they are so only through political means and institutions. Human rights have to be grounded in the commitment of individuals within a political community and therefore do not exist naturally, as something we come to possess through birth, but belong to the realm of the political, as only political institutions are able to put them into force.

This has turned out to be an extremely important insight which has laid bare the hypocrisy of the human rights discourse and has pointed to the way in which nation states are formed (based on exclusivity). However, Arendt's thoughts on the condition of statelessness also prohibit us from conceiving of refugees today as social actors who are able to become political agents. Following Arendt's argument, the stateless find themselves in the paradoxical situation of having the more fundamental human rights of freedom to speech, opinion and movement, without having any civic rights, due to their lack of belonging to a political community. A stateless person may enjoy more freedom of movement than a lawfully imprisoned criminal and more freedom of opinion than a citizen who is living under despotic rule. However, in such a case

> the prolongation of their lives is due to charity and not to right, for no law exists which could force the nations to feed them; their

freedom of movement, if they have it, gives them no right to residence which even the jailed criminal enjoys as a matter of course; and their freedom of opinion is a fool's freedom, for nothing they think matters anyhow.

(Arendt, 1958: 296)

For Arendt, the condition of being stateless thus has consequences beyond the loss of civic rights (although that is part of it) and entails the loss of 'the right to have rights'. The loss of the 'right to have rights' refers to the fundamental 'deprivation of a place in the world which makes opinions significant and actions effective' (Arendt, 1958: 296). This deprivation of a place in the world where action and speech can be meaningful is what it means to be 'absolutely rightless'. According to Arendt's understanding, without the right to action (to live in a world where our actions have meaning) and the right to speech (to be able to communicate meaningfully and formulate opinions) we are deprived of our humanity and hence are absolutely rightless. According to this account, political activism and agency of refugees, who have lost their 'right to have rights', is unthinkable. They would need to be the subject of a body of politics before they could qualify as a political subject in Arendtian terms.[3]

The second condition towards political agency for Arendt is freedom from necessity. In *The Human Condition* (1958/1998), Arendt argues that those who are not liberated from the struggle to sustain life appear as a threat to the public sphere since they would appropriate the realm of freedom to satisfy their natural needs, reducing politics to 'collective housekeeping'. To participate in a political struggle one must first be liberated from necessity, from the struggle to sustain and reproduce mere life in the obscurity of the private sphere. Arendt characterizes the mixing of these two kinds of struggle as anti-political. It is precisely freedom that allows humans to transcend life's necessities – the biological process that sustains life and attends to its needs – and to act. 'Action', as opposed to 'behaviour', is not guided by interests and is the activity which is done for its own sake (Arendt, 1998). Behaviour, on the other hand, is a mode of comportment that is totally preoccupied with securing life's necessities. In particular, while the activity of labour corresponds to the condition of life and the need to sustain and reproduce it, the activity of politics corresponds to the condition of plurality and the impulse to distinguish ourselves from others. This enables Arendt to distinguish between political, unpolitical and anti-political forms of struggle. In political struggle, individuals strive to distinguish themselves and thereby realize their humanity in a public sphere that is constituted by a community of equal and distinct persons. However, to participate in this higher form of struggle one must first be liberated from necessity, from the struggle to sustain and reproduce mere life in the obscurity of the private sphere. In grounding these two forms of struggle on the ontological conditions of life and plurality, Arendt is able to insist that they must be kept separate.

On the one hand, taking to heart Arendt's emphasis on the refugee as an important political figure, and of action as constituting the political, but on the other hand, also taking her arguments on the limitations of the political agency of the refugee seriously when reading the increasing refugee protests throughout Europe, our central question is: How can refugees' protests be considered political actions given on the one hand the fact that refugees lack even the fundamental 'right to have rights' and on the other hand given that they are not free, as Arendt argues, of 'life necessities'? We think that this cannot be answered by remaining strictly within an Arendtian theoretical frame of thinking, but can be answered by employing both the concept of the politics of Jacque Rancière as dissensus, and the notion of precarity as a basis for political action put forward by Judith Butler.

Politics as dissensus

Ranciere has argued that Arendt's aporetic analysis of human rights makes it difficult to account for how stateless people might claim the 'right to have rights'. Rancière, in contrast with Arendt, does not understand the political in terms of the disclosure of a common world from a struggle for distinction among co-equals (Schaap, 2011: 23). Reciprocal recognition of equality is not a precondition for politics as for Arendt. He has argued that we must conceptualize the refugees as a form of taking subject rather than a having subject. In order to understand this, we have to take a closer look at Rancière's notion of politics as dissensus.

Rancière distinguishes politics from policing. Very provocatively, he calls policing (*la police*) everything that would normally be called politics. According to him, we must distinguish politics 'from what normally goes by the name of politics and for which l propose to reserve the name policing' (1999: xiii). When Rancière refers to 'what normally goes by the name of politics', he means the actions of assemblies and parliaments, the decisions of courts, the work of politicians and all of the efforts of bureaucrats. He uses it to name any order of hierarchy. Thus it would be policy-making as well as cultural and economic arrangements.

It connotes the vertical organization of society, the dividing up and distribution of the various parts that makeup the social whole. Thus, for Rancière, that which keeps the power system whole, which provides a totalizing (without remainder and without exclusion) account of the population by assigning everyone a title and a role within the social edifice, is called *la police*. It is 'an order of bodies that defines the allocation of ways of doing, ways of being, and ways of saying, and sees that those bodies are assigned by name to a particular place and task: it is an order of the visible and the sayable' (Rancière, 1999: 29). By offering a totalizing account of the situation, *la police* precludes the possibility of politics.

Rather differently, then, politics is defined in Rancière's work as something that breaks the order of the sensible and *la police*: 'politics comes solely through interruption, the initial twist that institutes politics as the deployment of a wrong or of a fundamental dispute' (Rancière, 1999: 13). Politics only comes about when the order of *la police* is broken by dissensus. In the instance of dissensus, the order that is structured is forced to admit that it is not capable of totalizing the situation without remainder and without exclusion. In this way, the political moment is a radical egalitarian moment. Because this scene of dissensus occurs on the edge of what is thinkable, Rancière's conception of politics remains far from an institutionalization, or offering a codifiable form of improving our current paradigm. The moment of dissensus is something that was previously impossible to perceive, and thus the idea of prescribing this moment in a law or juridical practice is, according to him, an impossibility (Van Den Hemel, 2008).

A political moment is then a moment in which included subjects that do not belong (more specifically the refugees) act on the current situation – by showing that they 'didn't have the rights that they had' and 'had the rights that they didn't' (Rancière, 2004: 304). Clearly, when we think of the refugee movements, Rancière brings forward a theory that can offer grounds for action and change. The definitions of politics that he employs are based on radical disruption of the situation. This disruption is then the basis for affirmative change of the situation so that it can accommodate this disruption. This accommodation can change all possible aspects of a society.

By framing politics as dissensus, Rancière argues that it always ultimately calls into question the distinction between what is essentially political and what is not. Thus for him, equality is 'an imperative, a contested claim of politics itself, indeed perhaps the main object of politics, not just an implicit necessary condition of it'. Equality itself has no substantive content but only makes sense in the particular context of social inequality that it challenges. As Rancière points out, politics typically involves a struggle by subaltern groups to be seen and heard as speaking subjects within a social order that denies that they are qualified to participate in politics. On this account the subject of human rights emerges through political action and speech that seeks to verify the existence of those rights that are inscribed within the self-understanding of the political community. The political is constituted when those who are not qualified to participate in politics presume to act and speak as if they are.

According to Rancière, if human rights are reducible to the rights of citizenship, they are redundant: they are the rights of those who already have rights. On the other hand, if they attach to the human as such, independent of their membership of a political community, they amount to nothing: they are the 'mere derision of right', the 'rights of those who have no rights', the 'paradoxical rights of the private, poor, unpoliticized individual' (Rancière, 2004: 298). Clearly for Arendt, the human in the human rights has an ontological existence of mere or bare life that is deprived of politics, and that sets them apart from the citizen, while for Rancière the human is always already a politically contested figure, there seems to be no pre-politically ontological human in his understanding. This actually comes close to Butler's arguments about the framing of a human, where some humans are framed in ways in which their humanity is even contested – they are considered as lives not worth living, thus their deaths are not worth mourning. For Rancière, political exclusion is precisely this non-recognition of particular categories of people as subjects who are qualified to speak (Schaap, 2011: 30). This is what Butler calls misframing.

Contra Arendt, then, Rancière suggests that 'the Rights of Man are the rights of those who have not the rights they have and have the rights they have not' (Rancière, 2004: 302). According to this framework of understanding the political, the refugees by acting together in protest say two important things:

- 'we have the rights we have not' (we are free): which means that even though we are not supposed to, expected to, allowed to act, we do act. This is a political statement in the sense that it questions the very way in which the political is defined or staged, and also who is included and who is excluded. This is 'acting despite of'. This is then claiming or taking agency. It is a subversion of the lack of the 'right to have rights' (as right for speech and action). This is claiming that right, announcing it, acting on it. There is no addressee here in the sense that no one can grant that right; that right can only be enacted by a collective, and by enacting it, it creates that very right.

- 'we have not the rights we have' (we are equal): which means that according to different universal declarations to which most states are accountable, they have many civil rights and they know that they are supposed to have them but do not have them, and that is a political statement in the sense that it holds someone responsible for not granting them. This is 'acting for'. This is claiming some kind of 'victimhood' or vulnerability, and it has an addressee. Someone has to react to this call and forge measures for equality.

This second point is an interesting one because it relates not merely to the fact of the act of the protest (like the first point) but also to the grammar of it, to its demands, to its means and tools. By saying 'we have not the rights we have', the refugees are staging precarity as equality, as the basis of political action. This is different from a frame of victimhood in terms of humanitarian subjectivity, which says: 'Let's help them, because they are in need'. In order to understand the political potential of this framing, we have to make use of the concept of precarity as understood by Judith Butler.

Politics of precarity

Rancière, as we have argued, is able to account for the refugee as a political agent by merely staging dissensus and declaring both that 'they have the rights they have not' and that they 'have not the rights they have', and as such he challenges the Arendtian conception of the 'right to have rights' as a precondition for political action. He also criticizes her for separating the social from the political through the argument of 'political life must be free of life necessities', an argument which, he contends, reifies and naturalizes a division between who is qualified for politics and who is not (Rancière, 2003). Complementing his analysis, we think that the best answer to Arendt's separation between the concept of *zoë* and the *bios politicos* are the arguments that are put forward by Judith Butler, both in *Precarious Life* (2004) and in *Frames of War* (2009).

Butler argues for the recognition of precariousness as a shared condition of all human life (2004, 2009). Precariousness means living socially, the fact that one's life is always in the hands of another, constituting a social network of hands. This implies a dependency on people we know, barely know or know not at all. Such recognition of shared precariousness, according to her, introduces strong normative commitments of equality and invites a more robust universalizing of rights that seeks to address basic human needs for food, shelter, and other conditions for persisting and flourishing. As there is no life without the conditions of life that variably sustain life, these conditions are our shared political responsibility.

Avoiding analytic traps regarding the reliance of humanity on 'nature', Butler links the existential conception of 'precariousness' with a more specifically political notion of 'precarity'.[4] Precariousness is an existential condition: all human life is precarious since we depend on the care-taking of others ever since we are born. The concept of precarity, on the other hand, focuses on the differential distribution of precariousness among people. According to Butler, 'precarity designates that politically induced condition in which certain populations suffer from failing social and economic networks of support and become differentially exposed to injury, violence and death. Such populations are at heightened risk of disease, poverty, starvation, displacement, and of exposure to violence without protection' (Butler, 2009: 25). And it is, in her view, the differential allocation of precarity that forms the point of departure for both a rethinking of bodily ontology and for progressive or left politics in ways that continue to exceed and traverse the categories of identity.

Our capacity to respond to 'otherness' in general will depend in part on how the differential norm of the human is communicated through visual and discursive frames. There are ways of framing that will bring the human into view in its frailty and precariousness, that will allow us to stand for the value and dignity of human life, to react with outrage when lives are degraded or eviscerated without regard for their value as lives. Frames, as Judith Butler points out, are operations of power that occur on ontological, epistemological and ethical levels. They regulate the affective and ethical dispositions through which phenomena are not only understood but also constituted. Frames matter in terms of what is problematized and in what manner. They also matter on the level of who and what gets recognized as a subject, as part of a broader understanding of humanity, or as a life form that is worth protecting, producing thus a 'historically contingent ontology'. The question of the recognition of life, which Butler elaborates, begs the question of norms and normativity: what norms operate in producing certain subjects as 'recognizable' persons and make others more difficult to recognize?

Butler's critical inquiries into the category of 'life' and what it means to be human have led her to consider the conditions of vulnerability and precariousness as being central to a reformulation of subjectivity based on a social ontology. This idea of subjectivity opposes a self-sufficient and deliberative autonomous individual in the name of the differential distribution of precariousness, vulnerability and a most radical relationality that characterizes the life of an embodied subject. A different social ontology, she proposes, would have to start from the presumption that there is a shared, and at the same time an unequally distributed, condition of precarity that situates our political lives, establishing our dependence on political and social institutions to persist.

Staging dissensus through embodied protest

Both frameworks of dissensus and precarity should allow us to interpret the political actions of the refugee protest in Vienna. First, following Rancière, the refugees did not ask for 'some' rights that would allow them to enter and participate in the space of politics. They did not follow the hierarchical and vertical idea of politics, or what Rancière coined *la police*. Instead, the refugees appropriated the space that was seen as being inaccessible to them by staging dissensus. For them, politics was thus contesting the very core of politics, contesting the very meaning of it. Through their actions they also showed that *la police* is not able to totalize all of the social order, and assign all parts their place.

In line with Rancière, the refugees demonstrate on the one hand that 'they have the rights that they have not'. They demonstrate their freedom as speaking beings despite being deprived of legal personhood or citizenship. They occupy a public space, and a church, to draw attention to their participation within a society instead of remaining unseen and unheard. The political is thus constituted when those who are not qualified to participate in politics presume to act and speak as if they are. By acting as if they have the rights that they lack, the refugees actualize their political freedom and equality. Even if the public sphere has been defined through their exclusion, they act.

On the other hand, they demonstrate that 'they have not the rights that they have'. They do not enjoy the rights that they are supposed to have according to the various human rights treaties to which countries are signatory. By publicizing their political exclusion, they draw attention to their plight and the ways in which they are denied the same universal human rights from which the state claims to derive its legitimacy. The protesters in this movement came from different communities, had no common identity or language, at times not even a common standpoint. Yet their actions described above draw our attention to the commonalities of the movement, which lie in the means that were used to carry out the protest, with the body as the most crucial instrument. In other words, when you have no 'right to have rights', all you have is your body.

The march from Traiskirchen to Vienna was a demonstration of how far they could go with their bodies. Living in the camp was both the appropriation of public space through the presence of their bodies and a deliberate demonstration of the vulnerability of these bodies, who were sleeping there, eating and drinking in public, a demonstration which had its most striking and powerful moments during the eviction of the camp, when the fragility and vulnerability of everything that they had built up became manifest through its destruction. As the bodies were driven out of this public space, the body itself became the locus of the protest during the hunger strike. The refugees strategically employed their condition of precariousness to point at the precarity of their situation, as they intentionally provoked and increased the vulnerability of their bodies by putting it at risk of dying. However, their precarity did not only begin at the moment of entering hunger strike. The slogan of 'we are here, because you were there' points at the history of colonialism and ongoing economic exploitation, defended and pushed forward with political means (see also McGauran, 2010: 52). Thus their increased precariousness is not only the means but also the reason for their political struggle for the right to stay and work. This establishes their exposure and their precarity, the ways in which they depend on political and social institutions to persist.

In other words, contra Arendt, Butler would argue that it was only when those needs that are supposed to remain private came out into the day and night of the square, formed into image and discourse for the media, did it finally become possible to extend the space and time of the event with such tenacity to bring attention to the movement. This is not to romanticize the political means that these refugees had at hand. There is nothing poetic about the precarity of hunger strike. For decades now, refugees' forms of activism have been 'shaped by historical ties between destination and home country, different political opportunity structures and access to formal political citizenship' (McGauran, 2010: 52). Acknowledging this, we are led towards a critique of the social structures, in which auto-aggressive acts turn into (some) people's most effective means to be heard. The fatality of these structures is staged most evidently, as among all of the possibilities to enact active 'non-citizenship' it was these auto-agressive acts that were eventually effective in gaining the attention of the public authorities:

The horror and condemnation shown by the public against these destructive means of protest – also see the refugee protest in Germany, where people sewed their lips – misses a point – these forms of protest were effective in that they managed to do something, that years of political self-organization among refugees and supporting organisations had not achieved – to get the attention of people in power. That is because these acts adhered to the established rationality of how politics are made in the field of migration today. This rationality is to be condemned, and the ones who reproduce it. The horror lies not only with the fact that people sew their mouths in order to be heard, but that it is the only option they feel that is left to them – and that reality proves them right.

(taz_flüchtlingsproteste.doc)

Conclusion

In this chapter we have argued against the hegemony of the two main ways in which refugees as subjects have been framed, one in terms of a humanitarian subject and the other in terms of a 'dangerous' subject. The refugee protests that have taken place all over Europe in the last two decades have profoundly challenged both of these depoliticizing frames by casting the refugee unquestionably into the political sphere. Starting from this challenge put forward by the refugees themselves, we give additional voice to that frame and we read the refugee movements, focusing on the case of the Vienna protests which have started in 2012, as constituting political action.

While on the one hand Hannah Arendt has offered both her focus on the refugee as a central figure in political theory and her focus on action as constituting political agency, we failed to find within an Arendtian framework of thinking answers to the question of how political agency is constituted by refugees' protests and actions. We found this particularly difficult, given on the one hand her argument that bases political agency on the 'right to have rights' and makes unthinkable for a refugee who does not belong to a body politics to create political agency, and on the other hand her insistence on political action to be free of life necessities.

Arguing instead with Jacques Rancière and Judith Butler and applying their ideas of politics as dissensus and precarity to our reading, we challenged both of these claims and argued that the protest constitutes political action precisely because the refugees take agency in spite of their lack of the 'right to have rights', through the staging of their precarity and life necessities.

Notes

1. These accounts are based on the personal experience of one of the authors, who actively supported the refugee protest movement in Vienna. For a self-representative account of its steps and stages, see Refugee Protest Camp Vienna (2014).
2. The Austrian Freedom Party (Freiheitliche Partei Österreichs).
3. This strong conclusion is surprising given her political arguments regarding action, identity and freedom. In *The Human Condition* and in later works, Arendt repeatedly argued that a political status is defined, or even created, by actions. One does not acquire a political presence and then begins to act. It is through actions that one reveals oneself as a political agent and no one can be considered a political agent before acting. Arendt argues in her lecture entitled 'Freedom and politics' that 'men are free – as distinguished from their capacity to freedom as long as they act, neither before nor after; for to be free and to act are the same' (Arendt, 1960: 33). Freedom, she says, is a state of being that manifests itself in action. Similarly, identity is craved and constituted only through action (what we do is who we are). The subject is an emergent rather than an antecedent property of action.
4. For more on the use of these concepts by other authors, see Banki (2013); Harney (2013); and Levinson (2010).

Bibliography

Aas, K.F. (2011) 'Crimmigrant' bodies and bona fide travelers: Surveillance, citizenship and global governance, *Theoretical Criminology*, 15(3): 331–346.

Aas, K.F. and Bosworth, M. (2013) *The Borders of Punishment. Migration, Citizenship, and Social Exclusion*. New York: Oxford University Press.

Anderson, M. (1996) *Do No Harm: Supporting Local Capacities for Peace through Aid*, Cambridge MA: Local Capacities for Peace Project. The Collaborative for Development Action, Inc.

Arendt, H. (1958) *Origins of Totalitarianism*, 2nd edn. Cleveland and New York: The World Publishing Company.

Arendt, H. (1960) Freedom and politics: A lecture, *Chicago Review*, 14(1): 28–46.

Arendt, H. (1998) *The Human Condition*, 2nd edn. Chicago: University of Chicago Press.

Banki, S. (2013) *The Paradoxical Power of Precarity: Refugees and Homeland Activism*. Accessed on 9 June 2014 at http://refugeereview.wordpress.com/2013/08/07/the-paradoxical-power-of-precarity/.

Benhabib, S. (2004) ' "The right to have rights": Hannah Arendt on the contradictions of the nation-state'. *The Rights of Others: Aliens, Residents and Citizens*. Cambridge: Cambridge University Press.

Blitz, B.K. and Otero-Iglesias, M. (2011) Stateless by any other name: Refused asylum-seekers in the United Kingdom, *Journal of Ethnic and Migration Studies*, 37(4): 657–573.

Borren, M. (2008) Towards an Arendtian politics of in/visibility. On stateless refugees and undocumented aliens, *Ethical Perspectives: Journal of European Ethics Network*, 15(2): 213–237.

Bousfield, D. (2005) *The Logic of Sovereignty and the Agency of the Refugee: Recovering the Political from 'Bare Life'*, YCISS working paper no. 36. Toronto: York Centre for International and Security Studies.

Buonfino, A. (2004) Between Unity and Plurality: The Politicization and Securitization of the Discourse of Immigration in Europe. New Political Science 26 (1): 23–48.

Butler, J. (2004) *Precarious Life: The Power of Mourning and Violence*. New York: Verso.

Butler, J. (2009) *Frames of War. When Is Life Grievable?* London and New York: Verso.

Butler, J. (2011) 'Bodies in alliance and the politics of the street', European Institute for Progressive Cultural Politics, Vienna. Accessed on 9 June 2014 at http://eipcp.net/transversal/1011/butler/en.

Castañeda, H. (2010) 'Deportation deferred: "Illegality", visibility, and recognition in contemporary Germany'. In de Genova, N. and Peutz, N. (eds) *The Deportation Regime: Sovereignty, Space, and the Freedom of Movement*. Durham: Duke University Press.

Cissé, M. (1999) *Parole de sans papiers*. Paris: La Dispute.

Cotter, B.L. (2005) 'Hannah Arendt and the "right to have rights"'. In Lang, A.F. and Williams, J. (eds) *Hannah Arendt and International Relations: Readings across the Lines. Palgrave Macmillan History of International Thought*. New York: Palgrave Macmillan.

Fassin, D. (2007) Humanitarianism as a politics of life, *Public Culture*, 19(3): 499–520.

Fassin, D. (2012) *Humanitarian Reason: A Moral History of the Present*. Berkeley: University of California Press.

Harney, N. (2013) Precarity, affect and problem solving with mobile phones by asylum seekers, refugees and migrants in Naples, Italy, *Journal of Refugee Studies*, 26(4): 541–557.

Krause, M. (2008) Undocumented migrants: An Arendtian perspective, *European Journal of Political Theory*, 7: 331–348.

Levinson, H. (2010) *Refocusing the Refugee Regime: From Vagrancy to Value*, University of North Carolina at Charlotte. Accessed on 10 June 2014 at http://commons.pacificu.edu/cgi/viewcontent.cgi?article=1012&context=rescogitans.

Lammers, E. (1999) *Refugees, Gender, and Human Security: A Theoretical Introduction and Annotated Bibliography*. Netherlands: International Books.

Lavenex, S. (2001) *The Europeanisation of Refugee Policies: Between Human Rights and Internal Security*. Aldershot: Ashgate.

McGauran, K. (2010) 'We are here, because you were there. Migration activism in Europe'. In Bakondy, Vida; Ferfoglia, Simonetta; Janković, Jasmina; Kogoj, Cornelia; Ongan, Gamze; Pichler, Heinrich; Sircar, Ruby; Winter, Renée für die Initiative Minderheiten (eds) *Viel Glück! Migration Heute. Wien, Belgrad. Zagreb, Istanbul*. Wien: Mandelbaum Verlag.

Nyers, P. (2003) Abject cosmopolitanism: The politics of protection in the anti-deportation movement, *Third World Quarterly*. 24(6): 1069–1093.

Oman, N. (2010) Hannah Arendt's 'right to have rights': A philosophical context for human security, *Journal of Human Rights*, 9: 279–302.

Papastergiadis, N. (2006) The invasion complex: The abject other and spaces of violence, *Geografiska Annaler*, 88B(4): 429–442.

Pojmann, W. (ed.) (2008) *Migration and Activism in Europe since 1945*. New York: Palgrave Macmillan.

Pickering, S. (2001) Common sense and original deviancy: News discourses and asylum seekers in Australia, *Journal of Refugee Studies*, 14(2): 169–186.

Rajaram, P.K. (2002) Humanitarianism and representations of the refugee, *Journal of Refugee Studies*, 15(3): 247–264.

Rancière, J. (1999) *Disagreement: Politics and Philosophy*. Minneapolis: University of Minnesota Press.

Rancière, J. (2001) Ten thesis on politics, *Theory & Event*, 5: 3.

Ranciere, J. (2003) *The Philosopher and His Poor*. Durham, NC: Duke University Press.

Rancière, J. (2004) Who is the subject of the rights of man? *South Atlantic Quarterly*, 103(2/3): 297–310.

Refugee Protest Camp Vienna (2014). Accessed on 7 June 2014 at refugeecampvienna.noblogs.org.

Schaap, A. (2011) Enacting the right to have rights: Jacques Rancière's critique of Hannah Arendt, *European Journal of Political Theory*, 10: 22–45.

Shane Boyle, M. (2011) The criminalization of dissent. protest violence, activist performance, and the curious case of the VolxTheaterKarawane in Genoa, *TDR: The Drama Review*, 55(4): 113–127.

Soguk, N. (1999) *States and Strangers: Refugees and Displacements of Statecraft*. Minneapolis: University of Minnesota Press.

The VOICE Refugee Forum Germany (2014). Accessed on 9 June 2014 at http://thevoiceforum.org.

Van den Hemel, E. (2008) Included but not belonging. Badiou and Rancière on human rights, *Krisis Journal for Contemporary Philosophy*, 3: 16–30.

Wijzijnhier (2014). Accessed on 9 June 2014 at http://wijzijnhier.org.

Zetter, R. (1991) Labelling refugees: Forming and transforming a bureaucratic identity, *Journal of Refugee Studies*, 4(1): 39–62.

Zierer, B. (1998) Politische Flüchtlinge in österreichischen Printmedien (Political refugees in Austrian print media). Wien: Braumüller.

Index

Note: locators followed by 'n' refer to note numbers.

Printed and bound by CPI Group (UK) Ltd, Croydon, CR0 4YY